Preparing for Future Products of Biotechnology

Committee on Future Biotechnology Products and Opportunities to
Enhance Capabilities of the Biotechnology Regulatory System

Board on Life Sciences

Board on Agriculture and Natural Resources

Board on Chemical Sciences and Technology

Division on Earth and Life Studies

A Report of

The National Academies of
SCIENCES · ENGINEERING · MEDICINE

THE NATIONAL ACADEMIES PRESS
Washington, DC
www.nap.edu

THE NATIONAL ACADEMIES PRESS   **500 Fifth Street, NW**   **Washington, DC 20001**

This activity was supported by Contract No. EP-C-14-005 with the U.S. Environmental Protection Agency. Any opinions, findings, conclusions, or recommendations expressed in this publication do not necessarily reflect the views of any organization or agency that provided support for the project.

International Standard Book Number-13:   978-0-309-45205-2
International Standard Book Number-10:   0-309-45205-8
Digital Object Identifier: https://doi.org/10.17226/24605
Library of Congress Control Number: 2017940892

Additional copies of this publication are available for sale from the National Academies Press, 500 Fifth Street NW, Keck 360, Washington, DC 20001; (800) 624-6242 or (202) 334-3313; http://www.nap.edu.

Suggested citation: National Academies of Sciences, Engineering, and Medicine. 2017. *Preparing for Future Products of Biotechnology*. Washington, DC: The National Academies Press. doi: https://doi.org/10.17226/24605.

The National Academies of
SCIENCES · ENGINEERING · MEDICINE

Reports document the evidence-based consensus of an authoring committee of experts. Reports typically include findings, conclusions, and recommendations based on information gathered by the committee and committee deliberations. Reports are peer reviewed and are approved by the National Academies of Sciences, Engineering, and Medicine.

Proceedings chronicle the presentations and discussions at a workshop, symposium, or other convening event. The statements and opinions contained in proceedings are those of the participants and have not been endorsed by other participants, the planning committee, or the National Academies of Sciences, Engineering, and Medicine.

For information about other products and activities of the National Academies, please visit nationalacademies.org/whatwedo.

COMMITTEE ON FUTURE BIOTECHNOLOGY PRODUCTS AND OPPORTUNITIES TO ENHANCE CAPABILITIES OF THE BIOTECHNOLOGY REGULATORY SYSTEM

Chair

RICHARD M. MURRAY, NAE[1], California Institute of Technology, Pasadena

Members

RICHARD M. AMASINO, NAS[2], University of Wisconsin–Madison
STEVEN P. BRADBURY, Iowa State University, Ames
BARBARA J. EVANS, University of Houston Law Center, Houston, TX
STEVEN L. EVANS, Dow AgroSciences, Indianapolis, IN
FARREN ISAACS, Yale University, New Haven, CT
RICHARD A. JOHNSON, NAS[2], Global Helix LLC, Washington, DC (resigned August 2016)
MARTHA A. KREBS, The Pennsylvania State University, Philadelphia
JENNIFER KUZMA, North Carolina State University, Raleigh
MARY E. MAXON, Lawrence Berkeley National Laboratory, Emeryville, CA
RAUL F. MEDINA, Texas A&M University, College Station
DAVID REJESKI, Environmental Law Institute, Washington, DC
JEFFREY WOLT, Iowa State University, Ames

Staff

KARA N. LANEY, Study Director (from July 2016)
DOUGLAS FRIEDMAN, Study Director (through July 2016)
MARILEE K. SHELTON-DAVENPORT, Senior Program Officer
ANDREA HODGSON, Postdoctoral Fellow
JENNA BRISCOE, Senior Program Assistant
AANIKA SENN, Senior Program Assistant
FRANCES E. SHARPLES, Director, Board on Life Sciences
ROBIN A. SCHOEN, Director, Board on Agriculture and Natural Resources
TERESA FRYBERGER, Director, Board on Chemical Sciences and Technology

---

[1]National Academy of Engineering.
[2]National Academy of Sciences.

v

---

[1]National Academy of Engineering.
[2]National Academy of Sciences.

Preface

The Coordinated Framework for Regulation of Biotechnology is the U.S. government's mechanism for providing oversight for the diverse array of products of biotechnology that are being brought to market in a manner that protects health and the environment. In light of some of the recent advances in biotechnology, including in genome editing, gene drives, and synthetic biology, the federal government is in the process of making much needed updates to the Coordinated Framework, which had its last major revision in 1992. The task of this committee was to look into the future and describe the possible products of biotechnology that will arise over the next 5–10 years, as well as provide some insights that can help shape the capabilities within the agencies as they move forward. Given the rapid (and often unforeseen) advances of the *past* 5–10 years, it is clear that making accurate predictions of what will be possible is a difficult task, but some trends are clear: there will be a profusion of new products that will in many cases be very different in terms of their type, scope, and complexity, and the number of actors who will be able to contribute to biotechnology will be even more diverse as engineering biology becomes more accessible. At the same time, there is increased public awareness (and in some cases controversy), and the regulatory agencies are faced with the challenge of balancing the many competing interests from industry, society, government, and academia.

This report reflects the committee's deliberations regarding the future products of biotechnology that are likely to appear on the horizon, the challenges that the regulatory agencies might face, and the opportunities for enhancing the regulatory system to be prepared for what might be coming. The committee reached consensus on conclusions and recommendations that are based on extensive information gathering, committee discussions, and input from a wide variety of communities interested in biotechnology. The committee contained a diverse set of experts, including individuals with experience in natural sciences and engineering, law and public policy, social sciences, and risk assessment.

This report would not have been possible without the exceptional contributions of the National Academies of Sciences, Engineering, and Medicine staff members: Kara Laney, Marilee Shelton-Davenport, Andrea Hodgson, Jenna Briscoe, and Doug Friedman. Aanika Senn provided the committee with outstanding logistical support. We acknowledge gratefully all of their efforts.

A special thanks to my colleagues on the committee for their robust opinions and thoughtful discussions that helped shape the conclusions and recommendations that we reached. It was a pleasure to work with this outstanding group of experts.

Richard M. Murray, *Chair*
Committee on Future Biotechnology Products and
Opportunities to Enhance Capabilities of the Biotechnology
Regulatory System

Acknowledgments

This report is a product of the cooperation and contribution of many people. The members of the committee thank all the speakers who participated in the committee's meetings. (Appendix B contains a list of presentations to the committee.) The committee also wishes to thank all the agency officials who provided responses to the committee's request for information. (Appendix C contains a list of agencies sent the request.) In addition, the committee expresses its gratitude to Dr. Sarah Carter for her contributions to the committee's information-gathering process.

This report has been reviewed in draft form by individuals chosen for their diverse perspectives and technical expertise. The purpose of this independent review is to provide candid and critical comments that will assist the institution in making its published report as sound as possible and to ensure that the report meets institutional standards for objectivity, evidence, and responsiveness to the study charge. The review comments and draft manuscript remain confidential to protect the integrity of the process. We wish to thank the following individuals for their review of this report:

Lynn Bergeson, Bergeson & Campbell, PC
Yann Devos, European Food Safety Authority
Thomas Dietz, Michigan State University
Robert Friedman, J. Craig Venter Institute
Keith Hayes, Commonwealth Scientific and Industrial Research Organisation, Australia
Jennifer R. Holmgren, LanzaTech
Peter Barton Hutt, Covington & Burling LLP
Michael Jewett, Northwestern University
Sally Katzen, New York University School of Law
James Liao, Academia Sinica
Phil MacDonald, Canadian Food Inspection Agency
Joachim Messing, Waksman Institute of Microbiology
Warren Muir, Granite Research Institute
D. Warner North, NorthWorks
Reshma Shetty, Ginkgo Bioworks, Inc.

C. Neal Stewart, University of Tennessee
Daniel Voytas, University of Minnesota

Although reviewers listed above have provided many constructive comments and suggestions, they were not asked to endorse the report's conclusions or recommendations, nor did they see the final draft of the report before the release. The review of this report was overseen by **Lynn Goldman**, The George Washington University, and **Deborah Delmer**, University of California, Davis. They were responsible for making certain than an independent examination of this report was carried out in accordance with institutional procedures and that all review comments were carefully considered. Responsibility for the final content of this report rests entirely with the authoring committee and the institution.

Contents

SELECT ACRONYMS AND ABBREVIATIONS   xv

SUMMARY   1

1   INTRODUCTION AND CONTEXT   15
Impetus for the Study, 15
The Committee and Its Charge, 17
Information-Gathering Activities, 17
Organization of the Report, 18
Report Context and Scope, 19
References, 25

2   EMERGING TRENDS AND PRODUCTS OF BIOTECHNOLOGY   27
Setting the Stage: Understanding the Key Drivers for Future Biotechnology
    Products, 27
Future Biotechnology Products, 40
Summary and Conclusions, 57
References, 59

3   THE CURRENT BIOTECHNOLOGY REGULATORY SYSTEM   67
Overview of U.S. Regulatory System, 68
Consumer and Occupational Safety, 76
Environmental Protection, 90
Summary and Conclusions, 98
References, 102

4 UNDERSTANDING RISKS RELATED TO FUTURE BIOTECHNOLOGY PRODUCTS   107

Risks from Future Biotechnology Products: Similarities to the Past and Gaps Going Forward, 107

Existing Federal Capabilities, Expertise, and Capacity, 119

Summary and Conclusions, 132

References, 133

5 OPPORTUNITIES TO ENHANCE THE CAPABILITIES OF THE BIOTECHNOLOGY REGULATORY SYSTEM   137

Consistent, Efficient, and Effective Decision Making for Future Products of Biotechnology, 139

Technical Toolbox and Capabilities for Risk Assessment and Regulatory Science, 150

Summary and Conclusions, 164

References, 167

6 CONCLUSIONS AND RECOMMENDATIONS   171

Major Advances and New Types of Products, 172

Current Risk-Analysis System and Agency Authorities, 172

Future Products Under the Current Coordinated Framework, 174

Opportunities for Enhancement, 175

Recommendations, 176

References, 186

GLOSSARY   187

APPENDIXES

A BIOGRAPHICAL SKETCHES OF COMMITTEE MEMBERS   191

B AGENDAS OF INFORMATION-GATHERING SESSIONS   197

C REQUESTS FOR INFORMATION   203

D CONGRESSIONALLY DEFINED PRODUCT CATEGORIES THAT THE U.S. FOOD AND DRUG ADMINISTRATION REGULATES   209

Select Acronyms and Abbreviations

AHPA   Animal Health Protection Act
APHIS   Animal and Plant Health Inspection Service
ARS   Agricultural Research Service

BRAG   Biotechnology Risk Assessment Grants
Bt   *Bacillus thuringiensis*
BWG   Biotechnology Working Group

CAD   Computer-Aided Design
Cas9   CRISPR-Associated Protein 9
CDC   Centers for Disease Control and Prevention
CPSC   Consumer Product Safety Commission
CRISPR   Clustered Regularly Interspaced Short Palindromic Repeat
CSIRO   Commonwealth Scientific and Industrial Research Organisation

DARPA   Defense Advanced Research Projects Agency
DBTL   Design-Build-Test-Learn
DIYbio   Do-It-Yourself Biology
DNA   Deoxyribonucleic Acid
DOE   U.S. Department of Energy
dsDNA   Double-Stranded DNA
DSHEA   Dietary Supplement Health and Education Act

ELSI   Ethical, Legal, and Social Implications
EOP   Executive Office of the President
EPA   U.S. Environmental Protection Agency
ESA   Endangered Species Act
EUP   Experimental Use Permit

FACA	Federal Advisory Committee Act
FBI	Federal Bureau of Investigation
FDA	U.S. Food and Drug Administration
FDCA	Food, Drug, and Cosmetic Act
FIFRA	Federal Insecticide, Fungicide, and Rodenticide Act
FONSI	Finding of No Significant Impact
FTE	Full-Time Employee
FWS	U.S. Fish and Wildlife Service
GE	Genetically Engineered
GRAS	Generally Recognized as Safe
GRO	Genomically Recoded Organism
IARPA	Intelligence Advanced Research Projects Agency
IDE	Investigational Device Exemption
iGEM	International Genetically Engineered Machine
INAD	Investigational New Animal Drug
IND	Investigational New Drug
MAGE	Multiplex Automated Genome Engineering
MCAN	Microbial Commercial Activity Notice
NEPA	National Environmental Policy Act
NGO	Nongovernmental Organization
NIH	National Institutes of Health
NIST	National Institute of Standards and Technology
NMFS	National Marine Fisheries Service
NOI	Notice of Intent
NSF	National Science Foundation
ORD	Office of Research and Development
OSHA	Occupational Safety and Health Administration
PIP	Plant-Incorporated Protectant
PPA	Plant Protection Act
PPDC	Pesticide Program Dialogue Committee
rDNA	Recombinant DNA
RFI	Request for Information
RNA	Ribonucleic Acid
RNAi	RNA Interference
SAP	Scientific Advisory Panel
ssDNA	Single-Stranded DNA
TALEN	Transcription Activator-Like Effector Nuclease
TSCA	Toxic Substances Control Act

USDA　　　　U.S. Department of Agriculture
USGS　　　　U.S. Geological Survey

ZFN　　　　Zinc Finger Nuclease

Summary

In July 2015, the Office of Science and Technology Policy in the Executive Office of the President initiated an effort to modernize the U.S. regulatory system for biotechnology products consisting of three primary activities:

1. Development of an update to the Coordinated Framework for Regulation of Biotechnology (referred to hereafter as the Coordinated Framework) to clarify the roles and responsibilities of the agencies that regulate the products of biotechnology;
2. Formulation of a long-term strategy to ensure that the federal regulatory system is equipped to efficiently assess the risks, if any, associated with future products of biotechnology while supporting innovation, protecting health and the environment, promoting public confidence in the regulatory process, increasing transparency and predictability, and reducing unnecessary costs and burdens; and
3. Commission of an external, independent analysis of the future landscape of biotechnology products with a primary focus on potential new risks and risk-assessment frameworks.

With regard to the third item, the U.S. Environmental Protection Agency (EPA), the U.S. Food and Drug Administration (FDA), and the U.S. Department of Agriculture (USDA) were charged to

Commission an external, independent analysis of the future landscape of biotechnology products that will identify (1) potential new risks and frameworks for risk assessment and (2) areas in which the risks or lack of risks relating to the products of biotechnology are well understood. The intent of this review is to help inform future policy making. It is also anticipated that due to the rapid pace of change in this arena, an external analysis would be completed at least every 5 years.[1]

---

[1] Executive Office of the President. 2015. Memorandum for Heads of Food and Drug Administration, Environmental Protection Agency and Department of Agriculture. July 2. Available at https://obamawhitehouse.archives.gov/sites/default/files/microsites/ostp/modernizing_the_reg_system_for_biotech_products_memo_final.pdf. Accessed January 31, 2017.

BOX S-1
Statement of Task

An ad hoc committee of the National Academies of Sciences, Engineering, and Medicine will produce a report designed to answer the questions "What will the likely future products of biotechnology be over the next 5–10 years? What scientific capabilities, tools, and/or expertise may be needed by the regulatory agencies to ensure they make efficient and sound evaluations of the likely future products of biotechnology?" The committee will

- Describe the major advances and the potential new types of biotechnology products likely to emerge over the next 5–10 years.
- Describe the existing risk-analysis system for biotechnology products including, but perhaps not limited to, risk analyses developed and used by EPA, USDA, and FDA, and describe each agency's authorities as they pertain to the products of biotechnology.
- Determine whether potential future products could pose different types of risks relative to existing products and organisms. Where appropriate, identify areas in which the risks or lack of risks relating to the products of biotechnology are well understood.
- Indicate what scientific capabilities, tools, and expertise may be useful to the regulatory agencies to support oversight of potential future products of biotechnology.

Human drugs and medical devices will not be included in the purview of the study per a sponsor's request.

To accomplish this directive, the three regulatory agencies asked the National Academies of Sciences, Engineering, and Medicine to convene a committee of experts to conduct the study "Future Biotechnology Products and Opportunities to Enhance Capabilities of the Biotechnology Regulatory System." Committee members were selected because of the relevance of their experience and knowledge to the study's specific statement of task (Box S-1), and their appointments were approved by the President of the National Academy of Sciences in early 2016.

THE COMMITTEE'S PROCESS

To address its statement of task, the Committee on Future Biotechnology Products and Opportunities to Enhance Capabilities of the Biotechnology Regulatory System spent several months gathering information from a number of sources. It heard from 74 speakers over the course of three in-person meetings and eight webinars and received responses to a request for information from a dozen federal agencies. It also solicited statements from members of the public at its in-person meetings and accepted written comments through the duration of the study. The committee also made use of several recent National Academies studies related to future products of biotechnology, particularly *Industrialization of Biology: A Roadmap to Accelerate the Advanced Manufacturing of Chemicals*,[2] *Gene Drives on the Horizon: Advancing Science, Navigating Uncertainty, and Aligning Research with Public Values*,[3] and *Genetically Engineered Crops: Experiences and*

---

[2]NRC (National Research Council). 2015. Industrialization of Biology: A Roadmap to Accelerate the Advanced Manufacturing of Chemicals. Washington, DC: The National Academies Press.

[3]NASEM (National Academies of Sciences, Engineering, and Medicine). 2016. Gene Drives on the Horizon: Advancing Science, Navigating Uncertainty, and Aligning Research with Public Values. Washington, DC: The National Academies Press.

Prospects.[4] The committee reviewed these reports and reflected on their recommendations related to the Coordinated Framework, with the aim of understanding how those prior recommendations fit with the broader view of biotechnology products in this report and the opportunities to enhance the capabilities of the biotechnology regulatory system. For its purposes, the committee defined *biotechnology products* as products developed through genetic engineering or genome engineering (including products where the engineered DNA molecule is itself the "product," as in an engineered molecule used as a DNA information-storage medium) or the targeted or in vitro manipulation of genetic information of organisms, including plants, animals, and microbes. The term also covers some products produced by such plants, animals, microbes, and cell-free systems or products derived from all of the above.

FUTURE BIOTECHNOLOGY PRODUCTS

The committee was charged to describe biotechnology products likely to emerge in the next 5–10 years. The committee scanned the horizon for new products by inviting product developers to speak at the various meetings; reviewing submitted public comments; reading scientific literature, popular press reports, and patents; consulting previous reports by the National Academies; searching publicly available projects developed by international Genetically Engineered Machine teams;[5] and checking information available on regulatory agencies' websites and crowdfunding websites. It also made use of the Synthetic Biology Database[6] curated by the Woodrow Wilson Center. Based on this exercise, the committee anticipates that the scope, scale (number of products and variants thereof), and complexity of future biotechnology products may be substantially different from products developed as of 2016.

The committee grouped future products into three major classes: open-release products, contained products, and platforms. Table S-1 summarizes types of open-release products that the committee saw on the horizon, that is, plants, animals, microbes, and synthetic organisms that have been engineered for deliberate release in an open environment. The ability to sustain existence in the environment with little or no human intervention is a key change between existing products of biotechnology and some of the future ones anticipated in this class. Furthermore, the types of environments in which a product may persist are likely to become more diverse. Plants and insects may be designed to continue in low-management systems such as forests, pastures, and cityscapes; microbes may be developed to persist in those environments as well as in mines, waterways, and animal guts. The committee thought that future open-release products would be developed for familiar uses, such as agricultural crops, but would also likely be developed for uses such as cleaning up contaminated sites with engineered microbes, replacing animal-derived meat with meat cultured from animal cells, and controlling invasive species through gene drives.[7]

On the basis of its information-gathering efforts, the committee concluded that future biotechnology products that are produced in contained environments are more likely to be microbial based or synthetically based rather than based on an animal or plant host (Table S-2). Organisms of many genera are used in fermenters to produce commodity chemicals, fuels, specialty chemicals

---

[4]NASEM. 2016. Genetically Engineered Crops: Experiences and Prospects. Washington, DC: The National Academies Press.

[5]See team list for iGEM championship. Available at http://igem.org/Team_List?year=2016&name=Championship&division=igem. Accessed February 12, 2017.

[6]Synthetic Biology Products and Applications Inventory. Available at http://www.synbioproject.org/cpi. Accessed October 11, 2016.

[7]A *gene drive* is a system of biased inheritance in which the ability of a genetic element to pass from a parent to its offspring through sexual reproduction is enhanced. Thus, the result of a gene drive is the preferential increase of a specific *genotype*, the genetic makeup of an organism that determines a specific *phenotype* (trait), from one generation to the next, and potentially throughout the population.

TABLE S-1 Market Status of Products Designed for Open Release in the Environment[a]

	Product Description	On Market[b]	Under Development[c]	Early-Stage Concept
Plants and Plant Products	Bt crops with recombinant DNA[d] (rDNA)	✓		
	Herbicide-resistant crops with rDNA	✓	✓	
	Disease-resistant crops with rDNA	✓	✓	
	RNAi[e] modified crops	✓	✓✓✓	✓✓✓
	Fragrant moss		✓	
	Do-it-yourself glowing plants		✓	
	Genome-edited[f] crops		✓✓✓	✓✓✓
	Crops with CRISPR[g] knockouts		✓✓✓	✓✓✓
	Grasses for phytoremediation		✓	
	Plants as sentinels		✓	
	Crops with increased photosynthesis efficiency		✓	
	Ever-blooming plants			✓
	Nitrogen-fixing nonleguminous plants			✓
	Bioluminescent trees			✓
	Plants with gene drives for conservation purposes			✓
	Plants with gene drives for agricultural purposes			✓
Animals and Animal Products	Fluorescent zebra fish	✓		
	Sterile insects		✓	
	Genome-edited animals (e.g., polled cattle)		✓	✓
	Reduced-allergen goat's milk		✓	
	Landmine-detecting mice		✓	
	Animals revived from near extinction or extinction			✓
	Animals with gene drives for control of invasive mammals			✓
	Animals with gene drives for control of insect pests			✓
Microbes and Microbial Products	Biosensors/bioreporters		✓	
	Bioremediation		✓	
	Engineered algal strains		✓✓✓	
	Nitrogen-fixing symbionts		✓	
	Probiotics			✓
	Genomically engineered microbial communities			✓✓✓
	Biomining/bioleaching			✓✓✓
Synthetic Organisms/ Nucleic Acids	Cell-free products		✓	
	DNA barcodes to track products	✓	✓	
	RNA-based spray for insect-pest control		✓	
	Genomically recoded organisms			✓
	Biological/mechanical hybrid biosensors		✓	✓

✓✓✓ = an area the committee has identified as having high growth potential.

[a]The table reflects the market status of products at the time the committee was writing the report.

[b]"On Market" is equivalent to "in use"; thus, products that have received regulatory approval but are not in use were not considered by the committee to be "On Market."

[c]"Under development" spans products from the prototype stage to field trials.

[d]*Recombinant DNA* is a novel DNA sequence created by joining DNA molecules that are not found together in nature.

[e]*RNAi or RNA interference* is a natural mechanism found in nearly all organisms in which the levels of transcripts are reduced or suppressed and can be exploited with biotechnology to modify an organism.

[f]*Genome editing* is a specific modification of the DNA of an organism to create mutations or introduce new alleles or new genes.

[g]*CRISPR or clustered regularly interspaced short palindromic repeat* is a naturally occurring mechanism of immunity to viruses found in bacteria that involves identification and degradation of foreign DNA. This natural mechanism has been manipulated by researchers to develop genome-editing techniques.

TABLE S-2 Market Status of Contained Products[a]

	Product Description	On Market[b]	Under Development[c]	Early-Stage Concept
Animals/Plants and Animal/Plant Products	Transgenic laboratory animals (mini-swine, mice, rats, dogs)	✓	✓✓✓	
	Genetically engineered salmon grown in land-based facilities	✓		
	Animal cell culture–derived products (e.g., cowless leather and cowless meat)		✓✓✓	✓✓✓
	Polymers produced by plants for industrial use		✓	
	Greenhouse crops with CRISPR knockouts		✓	
Microbes and Microbial Products	Industrial enzymes	✓	✓	✓
	Biobased chemicals to replace fossil fuel feedstocks	✓	✓	✓
	Bioluminescent microbes for home and landscape uses		✓	✓
	Yeast-derived molecules to create products (e.g., vanillin, stevia, saffron, egg whites, milk protein, gelatin)	✓	✓✓✓	✓✓✓
	Synthetic silk		✓	
	Bacterium-derived antimicrobials		✓	
	Genomically engineered bacterial strains for fermentation-based products		✓✓✓	
	Gas-phase microbial systems		✓	
	Algae-derived products (e.g., substitute for shark fins and shrimp, biofuels, ethylene)		✓✓✓	✓✓✓
	Probiotics			✓
	Leaching/metal recycling organisms			✓
Synthetic Organisms/ Nucleic Acids	Organ-on-a-chip		✓	
	V. natriegens platform	✓	✓	
	Genomically recoded organisms		✓✓✓	✓✓✓
	Cell-free expression systems		✓✓✓	✓✓✓
	Biological–mechanical hybrid biosensor		✓	✓
	Implantable biosensors		✓	✓

✓✓✓ = an area the committee has identified as having high growth potential.

[a]The table reflects the market status of products at the time the committee was writing the report.

[b]"On Market" is equivalent to "in use"; thus, products that have received regulatory approval but are not in use were not considered by the committee to be "On Market."

[c]"Under development" spans products from the prototype stage to field trials.

or intermediates, enzymes, polymers, food additives, and flavors. When considering the laboratory as a contained environment, many examples of transgenic animals from vendors are widely used today for research and development. Because performing biotechnology in contained environments allows higher control over the choice of host organism, systems with advanced molecular toolboxes are already in high use.

Biotechnology platforms are tools that are used in the creation of other biotechnology products. They include products that are traditionally characterized as "wet lab," such as DNA/RNA, enzymes, vectors, cloning kits, cells, library prep kits, and sequencing prep kits, and products that are "dry lab," such as vector drawing software, computer-aided design software, primer calculation software, and informatics tools. These two categories continue to meld as newer approaches are published or commercialized.

There are a variety of technical, economic, and social trends that are driving and that will continue to drive the types of biotechnology products developed in the next decade. Technical and economic trends in the biological sciences and biological engineering are accelerating the rate at which new product ideas are formulated and the number of actors who are involved in product development. With regard to social trends, it was evident to the committee through its information-gathering activities and the mechanisms for public comment that there are many competing interests, risks, and benefits regarding future biotechnology products; it was also clear that the United States and international regulatory systems will need to achieve a balance among these competing aspects when considering how to manage the development and use of new biotechnology products. Many sectors of society have concerns over the safety and ethics of various biotechnologies, whereas others see prospects for biotechnology to address challenging social and environmental issues. Biotechnology products that are on the horizon are likely to generate substantial public debate. For example, gene-drive technology, for which there have already been numerous studies and reports regarding its use, is a technological advance that will increase the amount of public debate and for which society will have to take a balanced approach among the interested and affected parties, developers, and scientists.

THE BIOTECHNOLOGY REGULATORY PROCESS
AND THE COORDINATED FRAMEWORK

The committee was asked to describe the existing risk-analysis system for biotechnology products and to describe each agency's authorities as they pertain to the products of biotechnology. In order to carry out these portions of its statement of task, the committee reviewed the regulatory authorities that apply to biotechnology products.

The committee found that the Coordinated Framework appears to have considerable flexibility in statutory authority to cover a wide range of biotechnology products. In some cases, however, the jurisdictions of EPA, FDA, and USDA are defined in ways that may leave gaps or redundancies in regulatory oversight. Even when jurisdiction exists, the available legal authorities may not be ideally tailored to new and emerging biotechnology products. Furthermore, agencies other than EPA, FDA, and USDA will likely have responsibilities to regulate some future biotechnology products, and their roles are not well specified in the Coordinated Framework.

Despite the flexibility of the Coordinated Framework to cover a wide range of biotechnology products, the committee also found that the existing biotechnology regulatory system is complex and could be considered to appear fragmented, resulting in a system that is difficult for product developers—including individuals, nontraditional organizations, and small enterprises—as well as consumers, product users, and interested members of the public to navigate. This complexity can cause uncertainty and a lack of predictability for developers of future biotechnology products and creates the potential for loss of public confidence in oversight of future biotechnology products.

The increased rate of new product ideas means that the types and number of biotechnology products in the next 5–10 years may be significantly larger than the current rate of product introduction. EPA, FDA, USDA, and other relevant agencies will need to be prepared for this potential increase, including finding effective means of evaluation that maintains public safety, protects the environment, and satisfies the statutory requirements appropriate for each agency. The increased number of actors who are involved in product development means that the regulatory agencies will need to be prepared to provide information regarding the regulatory process to groups that may have little familiarity with the Coordinated Framework. This group of actors may include small- and medium-sized enterprises, do-it-yourself (DIY) bioengineers, or developers supported by crowdfunded activities with direct-to-consumer distribution models and the potential for domestic manufacturing.

UNDERSTANDING RISKS RELATED TO FUTURE BIOTECHNOLOGY PRODUCTS

The committee was asked to determine whether future products could pose different types of risks relative to existing products and organisms. In all the types of products summarized above, advances in biotechnology are leading to products that involve the transformation of less familiar host organisms, have multiple engineered pathways, are comprised of DNA from multiple organisms, or are made from entirely synthetic DNA. Such products may have few or no comparators[8] to existing nonbiotechnology products, which function as the baseline of comparison in current regulatory risk assessments of biotechnology products. Figure S-1 summarizes the progression in terms of complexity and novelty that the committee thought was likely in future biotechnology products over the next 5–10 years. Products that fit in column A are those similar to existing biotechnology products evaluated under the existing Coordinated Framework and for which current methods of risk assessment can be applied. Examples include new genetically engineered crops and fermentation-based production of small molecules, enzymes, or other biochemicals. Products described by column B are those that represent an expansion of the familiar set of organismal hosts and genetic pathways, for which there are few comparators but nonetheless well-established approaches to assessing risk. Examples include animal cell culture–derived products (such as cowless meat or leather) and plants for bioremediation, decoration, or other environmental or consumer use. Products in column C are those that are currently at the forefront of research activities, where the use of rapid design-build-test-learn cycles allows much more complex designs of genetic pathways in a wider variety of host organisms, but which also represent more sophisticated uses of products, such as open release into the environment of organisms intended to modify populations of natural organisms. Examples include genetically engineered mosquitoes for fighting malaria or the Zika virus, genomically engineered microorganisms, and implantable biosensors. Such products are on the horizon, but at the time the committee was writing its report, most had not yet entered the biotechnology regulatory system. The few that had entered the system had few or no nonbiotechnology products to which they could be compared, and, as they were first-of-their-kind products, no previous biotechnology product had established a path to follow through the regulatory system. Finally, products in column D represent those in which multiple organisms may be used in complex microbial communities, such as microbiome engineering and synthetic consortia for bioremediation or biomining applications. These products also have no comparators (or the relevance of potential comparators is ambiguous) and no established regulatory path.

For future biotechnology products in all degrees of complexity and novelty, the committee considered the risk-assessment endpoints related to human health or environmental outcomes, such as illness, injury, death, or loss of ecosystem function. It concluded that the endpoints are not new compared with those that have been identified for existing biotechnology products, but the intermediate steps along the paths to those endpoints have the potential to be more complex, more ambiguous, and less well characterized. In addition, the committee found that the scope, scale, complexity, and tempo of biotechnology products that are likely to enter the regulatory system in the next 5–10 years have the potential to critically stress the regulatory agencies, both in terms of capacity and expertise. Furthermore, many early-stage developers of biotechnology products or biological technology that may lead to products do not currently consider regulatory perspectives or future requirements during technology (and sometimes product) development, which has the potential to complicate the evaluation of risks associated with the release of future biotechnology products. It will clearly be important for EPA, FDA, USDA, and other agencies relevant to the future regulation of biotechnology products to maintain an assessment of the scope of these products and be prepared to evaluate them as they are submitted for regulatory assessment.

---

[8]The term *comparator* refers to a known nonbiotechnology organism that is similar to the engineered organism except for the engineered trait.

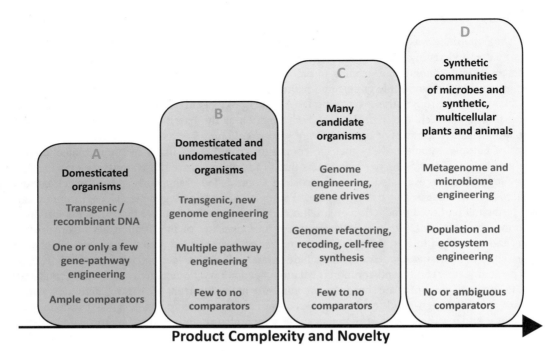

Product Complexity and Novelty

FIGURE S-1 Characteristics of future biotechnology products, organized by similar levels of complexity in terms of types and number of organisms, genes and traits, and comparators involved.
NOTE: Products of biotechnology can be conceptualized as fitting into the depicted columns with the indicated characteristics, moving toward column D as a product increases in complexity and likelihood of providing new challenges for risk assessment.

OPPORTUNITIES FOR ENHANCEMENT OF THE BIOTECHNOLOGY REGULATORY SYSTEM

A major task of the committee was to indicate what scientific capabilities, tools, and expertise may be useful to the regulatory agencies to support oversight of future products of biotechnology. The committee requested information from federal agencies regarding current investments in regulatory science.[9]

At a high level, the committee found that there are existing frameworks, tools, and processes for risk analyses and public engagement that can be used to address the many issues that are likely to arise in future biotechnology products in a way that balances competing issues and concerns. However, given the profusion of biotechnology products that are on the horizon, there is a risk that the capacity of the regulatory agencies may not be able to efficiently provide the quantity and quality of risk assessments that will be needed. An important approach for dealing with an increase in the products of biotechnology will be the increased use of stratified approaches to regulation, where new and potentially more complex risk-analysis methods will need to be developed for some

---

[9]As discussed in Chapter 4, on the basis of definitions provided by FDA and the Society for Risk Analysis, the committee understood regulatory science to involve developing and implementing risk-analysis methods and maximizing the utility of risk analyses to inform regulatory decisions for biotechnology products, consistent with human health and environmental risk–benefit standards provided in relevant statutes.

products, while established risk-analysis methods can be applied or modified to address products that are *familiar* or that require *less complex* risk analysis. With this approach, new risk-analysis methods are focused on products with *less familiar* characteristics and/or *more complex* risk pathways. Multiple criteria are usually embedded within risk analyses to ascertain if an estimated level of risk is consistent with the risk-management goals established during the problem-formulation phase of a risk assessment. In some cases, additional risk analyses may be needed to refine risk estimates, to evaluate risk-mitigation measures, or both. In order to implement the appropriate rigor of risk analyses for new biotechnology products, it will be necessary to establish scientifically rigorous criteria based on factors affecting the perception of risk, the degree of uncertainty, and the magnitude of risk and nature of potential risks.

To help articulate what capabilities, tools, and expertise might be useful to meet these objectives, the committee created a conceptual map for decision making aimed to assess and manage product risk, streamline regulation requirements, and increase transparency, as shown in Figure S-2.

As envisioned by the committee, a single point of entry (illustrated in Figure S-2) could be used by a product developer to evaluate whether the intended use of the product is regulated under

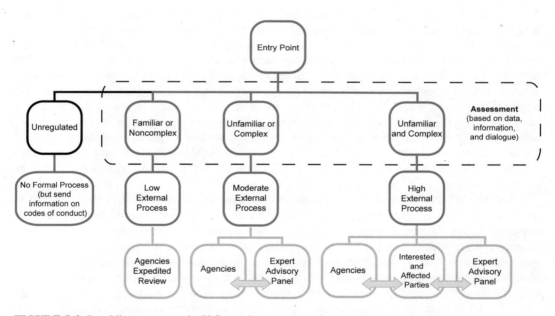

FIGURE S-2 Providing access to the U.S. regulatory system through a single point of entry.
NOTES: Potential product developers and interested parties would begin by going to an entry point and providing characteristics of the intended product and its use pattern. If the product does not fall under a federal statute, the developer would be notified that the product is not federally regulated. If the product is regulated, the appropriate agency or agencies would be identified for the developer. An evaluation of the product's familiarity to regulatory agencies and its complexity in terms of risk analyses as compared to existing biotechnology products would be ascertained (red bins). Depending on the product's familiarity and the complexity of its risk analysis, a different set of risk-analysis processes would be employed (blue boxes). For products that are *familiar* to the regulatory agencies and are *not complex*, a more expedited process could be used under the assumption that relevant risk-analysis processes are well established. For products that are *less familiar*, *more complex*, or *less familiar and more complex*, increasingly unique risk-analysis processes (that incorporate additional external input) may need to be established.

a given statute and provide a determination of whether the product is *familiar and not complex*, is *unfamiliar or complex*, or is *unfamiliar and complex* compared to existing biotechnology products. Once a determination has been made, the appropriate processes within the relevant agency (or agencies) would be used to provide the necessary risk analysis to support a regulatory decision. For products that are *familiar and noncomplex*, an expedited process might be used (for example, a notification process). For products that are determined to be *unfamiliar or complex* or *unfamiliar and complex*, new human health and ecological risk-analysis methods might be needed to inform a regulatory decision. A desirable feature of an integrated, stratified approach to regulatory oversight is that over time product types originally placed in the *unfamiliar or complex* bin or the *unfamiliar and complex* bin would "move" to a bin of less complexity or more familiarity based on experience gained in evaluating additional products in a category.

SUMMARY CONCLUSIONS

On the basis of its assessment of the trends in biotechnology, the likely products of biotechnology in the next 5–10 years, and the current authorities and capabilities of the regulatory agencies, the committee identified a set of broad themes regarding future opportunities for enhancement of the U.S. biotechnology regulatory system.

The bioeconomy is growing rapidly and the U.S. regulatory system needs to provide a balanced approach for consideration of the many competing interests in the face of this expansion. The competing interests and concerns articulated by the Executive Office of the President include supporting innovation, protecting human health, preserving biodiversity, reducing negative environment effects, promoting public confidence in the regulatory process, increasing transparency and predictability in the regulatory process, reducing unnecessary costs and burdens, making use of new tools from a broad range of disciplines, and interacting with the global economy. The pipeline of biotechnology products likely to emerge over the next decade probably will result in disruptive innovations and significant societal impacts; a carefully balanced, coordinated approach toward future biotechnology products that incorporates input from stakeholders—including interested and affected parties, relevant federal agencies, and nontraditional product developers—will be required.

The profusion of biotechnology products over the next 5–10 years has the potential to overwhelm the U.S. regulatory system, which may be exacerbated by a disconnect between research in regulatory science and expected uses of future biotechnology products. The number and complexity of products, new pathways to risk-assessment endpoints, large range of types of products (for example, those for open release in the environment or marketed as direct-to-consumer), new actors (including DIY bioengineers, small- and medium-sized enterprises, and crowdfunders), and complex alignment of potential future products with agency authorities are likely to change rapidly as biotechnology advances. A disconnect between research in regulatory science and its use in biotechnology research and product development creates a situation in which new products may be conceived and designed without sufficient consideration of regulatory requirements, which can lead to surprises and delays late in the development cycle. The update to the Coordinated Framework[10] and the *National Strategy for Modernizing the Regulatory System for Biotechnology Products*,[11] recently released by the Executive Office of the President, provide

---

[10]Executive Office of the President. 2017. Modernizing the Regulatory System for Biotechnology Products: An Update to the Coordinated Framework for the Regulation of Biotechnology. Available at https://obamawhitehouse.archives.gov/sites/default/files/microsites/ostp/2017_coordinated_framework_update.pdf. Accessed January 30, 2017.

[11]*National Strategy for Modernizing the Regulatory System for Biotechnology Products* is available at https://obamawhitehouse.archives.gov/sites/default/files/microsites/ostp/biotech_national_strategy_final.pdf. Accessed January 31, 2017.

an excellent starting point for addressing the products that will appear in the next 5–10 years. But additional investments are needed to be prepared for the subsequent generation of products that are on the horizon and to ensure that there is a consistent, efficient, and effective decision-making framework that continues to balance innovation and safety.

Regulators will face difficult challenges as they grapple with a broad array of new types of biotechnology products—for example, cosmetics, toys, pets, and office supplies—that go beyond contained industrial uses and traditional environmental release (for example, *Bt* or herbicide-resistant crops). The diversity of biotechnology products anticipated over the next decade confronts consumer- and occupational-safety regulators with two related challenges:

1. To find jurisdiction under existing statutes to regulate all the products that may pose risks to consumers and
2. To utilize the best available risk-analysis tools consistent with agency authorities to provide nuanced oversight that protects consumers while fostering beneficial innovation.

Existing statutes offer promising pathways to meet these challenges, although there may be cases when a novel product falls outside the jurisdiction of EPA, FDA, or USDA and is either in a jurisdictional gap (where no regulator has authority to address potential safety concerns) or under the jurisdiction of another agency, such as the Consumer Product Safety Commission, that has fewer statutory authorities and capabilities to conduct rigorous and timely risk analysis. For this reason, EPA, FDA, and USDA may at times need to make use of the flexibility available under their statutes to minimize gaps in jurisdiction and to position novel products under the statutory framework most suited to each product's characteristics and level of risk.

The safe use of new biotechnology products requires rigorous, predictable, and transparent risk-analysis processes whose comprehensiveness, depth, and throughput mirror the scope, scale, complexity, and tempo of future biotechnology applications. Regulatory oversight that is unnecessarily complex runs the risk of driving an "imitate not innovate" mentality and may not scale to match the pace of biotechnology innovation. Building on the approach outlined in the *National Strategy*, the committee believes that the advancement of existing risk-analysis methodologies within an easily accessible, participatory governance framework can establish an oversight process that matches the scope, scale, complexity, and tempo of future technological developments and increases public confidence in the safety of products entering the marketplace.

In addition to the conclusions and recommendations from this report, **EPA, FDA, USDA, and other agencies involved in regulation of future biotechnology products would benefit from adopting recommendations made by previous National Academies' committees related to future products of biotechnology, which are consistent with the findings and recommendations in this report.** Given the assessments of some future biotechnology products and the role of the regulatory system, many of the recommendations of previous National Academies' committees are directly relevant and should be considered when taking actions to enhance the capabilities of the U.S. biotechnology regulatory system.

SUMMARY RECOMMENDATIONS

On the basis of its conclusions, the committee developed a number of detailed recommendations regarding actions that can be taken to enhance the capabilities of the biotechnology regulatory system in order to be prepared for anticipated future products of biotechnology.

Recommendation 1: EPA, FDA, USDA, and other agencies involved in regulation of future biotechnology products should increase scientific capabilities, tools, expertise, and horizon scanning in key areas of expected growth of biotechnology, including natural, regulatory, and social sciences.

The information gathered by the committee indicates a substantial new set of technologies that are being brought to bear in future products and the agencies should continue to maintain their scientific capabilities across a broad range of disciplines. Example priority areas, discussed in more detail in the body of the report, including areas such as comparators, off-target gene effects, and phenotypic characterization; genetic fitness, genetic stability, and horizontal gene transfer; impacts on nontarget organisms; control of organismal traits; modeling (including risk-analysis approaches under uncertainty) and life-cycle analyses; monitoring and surveillance; and economic and social costs and benefits.

- Recommendation 1-1: Regulatory agencies should build and maintain the capacity to rapidly triage products entering the regulatory system that resemble existing products with a history of characterization and use, thus reducing the time and effort required for regulatory decision making, and they should be prepared to focus questions on identifying new pathways to risk-assessment endpoints associated with products that are unfamiliar and that require more complex risk assessments.
- Recommendation 1-2: In order to inform the regulatory process, federal agencies should build capacity to scan the horizon continuously for new products and processes that could present novel risk pathways, develop new approaches to assess and address more complex risk pathways, and implement mechanisms for keeping regulators aware of the emerging technologies they have to deal with.
- Recommendation 1-3: EPA, FDA, USDA, and other relevant federal agencies should work together to (a) pilot new approaches for problem formulation and uncertainty characterization in ecological risk assessments, with peer review and public participation, on open-release products expected during the next 5 years; (b) formulate risk–benefit assessment approaches for future products, with particular emphasis on future biotechnology products with unfamiliar functions and open-release biotechnology products; and (c) pool skills and expertise across the government as needed on first-of-a-kind risk–benefit cases.
- Recommendation 1-4: EPA, FDA, USDA, and other relevant federal agencies should create a precompetitive or preregulatory review "data commons" that provides data, scientific evidence, and scientific and market experience for product developers.
- Recommendation 1-5: Consistent with the goals and guidance stated by the Office of Science and Technology Policy in the Executive Office of the President in its July 2015 memo, the Biotechnology Working Group should implement a more permanent, coordinated mechanism to measure progress against and periodically review federal agencies' scientific capabilities, tools, expertise, and horizon scanning as they apply to the profusion of future biotechnology products.

Recommendation 2: EPA, FDA, and USDA should increase their use of pilot projects to advance understanding and use of ecological risk assessments and benefit analyses for future biotechnology products that are unfamiliar and complex and to prototype new approaches for iterative risk analyses that incorporate external peer review and public participation.

The rate of technology development in the biological sciences and engineering will create a situation in which many new types of products will be developed in the next 5–10 years. In order to handle the scope and complexity of future biotechnology applications, the regulatory agencies should make use of pilot products to identify ways to improve the comprehensiveness, effectiveness, and throughput of the regulatory process.

- Recommendation 2-1: Regulatory agencies should create pilot projects for more iterative processes for risk assessments that span development cycles for future biotechnology products as they move from laboratory scale to field or prototype scale to full-scale operation.
- Recommendation 2-2: Government agencies should pilot advances in ecological risk assessments and benefit analyses for open-release products expected in the next 5–10 years, with external, independent peer review and public participation.
- Recommendation 2-3: Government agencies should initiate pilot projects to develop probabilistic estimates of risks for current products as a means to compare the likelihood of adverse effects of future biotechnology products to existing biotechnology and nonbiotechnology alternatives.
- Recommendation 2-4: Regulatory agencies should make use of pilot projects to explore new methods of outreach to the public and developer community as a means of horizon scanning, assessing need areas for capability growth, and improving understanding of the regulatory process.
- Recommendation 2-5: EPA, FDA, and USDA should engage with federal and state consumer- and occupational-safety regulators that may confront new biotechnology products in the next 5–10 years and make use of pilot projects, interagency collaborations, shared data resources, and scientific tools to pilot new approaches for risk assessment that ensure consumer and occupational safety of new biotechnology products, particularly those that may involve novel financing mechanisms, means of production, or distribution pathways.

Recommendation 3: The National Science Foundation, the U.S. Department of Defense, the U.S. Department of Energy, the National Institute of Standards and Technology, and other agencies that fund biotechnology research with the potential to lead to new biotechnology products should increase their investments in regulatory science and link research and education activities to regulatory-science activities.

Increased investments in regulatory science will be needed to align desired science advancements with existing and anticipated regulatory requirements. It will be valuable for developers of biotechnology to incorporate regulatory perspectives earlier in the product and technology development process, and the research funding agencies can help enhance the regulatory system by increasing the awareness of regulatory science at an early stage.

- Recommendation 3-1: The federal government should develop and implement a long-term strategy for risk analysis of future biotechnology products, focused on identifying and prioritizing key risks for unfamiliar and more complex biotechnology products, and work to establish appropriate federal funding levels for sustained, multiyear research to develop the necessary advances in regulatory science.
- Recommendation 3-2: Federal agencies that fund early-stage biotechnology-related research and regulatory agencies should provide support to academic, industry, and government researchers to close gaps and provide linkages to market-path requirements for regulatory success.

- Recommendation 3-3: Government agencies that fund biotechnology development, working together with regulatory agencies and each other, should also invest in new methods of understanding the ethical, legal, and social implications associated with future biotechnology products.
- Recommendation 3-4: Government agencies with an educational mission, including those that support scientific training, should identify and fund activities that increase awareness and knowledge of the regulatory system in courses and educational materials for students whose research will lead to advances in biotechnology products.

Introduction and Context

Since Cohen and colleagues described recombinant-DNA (rDNA) techniques in their seminal 1973 publication (Cohen et al., 1973), humans have been able to directly manipulate gene sequences in organisms. The manipulation of DNA led to the development of new products; early examples include synthetic human insulin and virus-resistant squash. In the United States, it also led to the development of a new regulatory framework to oversee the introduction of these products into commerce and into the environment.

The Coordinated Framework for Regulation of Biotechnology (hereafter referred to as the Coordinated Framework) was finalized in 1986, a little more than a decade after rDNA techniques had first been used successfully and at a time when few products from such techniques were in commercial use. It was the U.S. government's policy "for ensuring the safety of biotechnology research and products" (OSTP, 1986:23303). In the notice announcing the policy, the Office of Science and Technology Policy (in the Executive Office of the President) stated that "[e]xisting statutes provide a basic network of agency jurisdiction over both research and products; this network forms the basis of this coordinated framework and helps assure reasonable safeguards for the public" (OSTP, 1986:23303). The existing statutes pertained to the U.S. Environmental Protection Agency (EPA), the U.S. Food and Drug Administration (FDA), and the U.S. Department of Agriculture (USDA), and the authorities granted under those statutes provided regulatory oversight jurisdiction for one or more of those federal agencies for all existing or foreseen biotechnology products at the time.

IMPETUS FOR THE STUDY

Between 1973 and 2016, the ways to manipulate DNA to endow new characteristics in an organism (that is, *biotechnology*) have advanced. *Genetic engineering*—the introduction or change of DNA, RNA, or proteins by human manipulation to effect a change in an organism's genome or epigenome—originally relied on the use of a second organism (often a bacterium) as a vector to introduce a desired genetic change into the organism of interest. However, biolistic particle delivery—also known as the gene gun—was developed in the 1980s and can insert genetic material into the organism of interest without the use of a vector organism. *Genome engineering* employs a direct and precise approach to whole-genome design and mutagenesis to enable a rapid and controlled exploration of an organism's phenotype landscape. Advances in genome engineering are being fueled by two prevailing approaches: genome synthesis and genome editing. *Whole-genome synthesis*, which combines *de novo* DNA synthesis, large-scale DNA assembly, transplantation, and recombination, permits *de novo* construction of user-defined double-stranded DNA throughout the whole genome. *Genome-editing* techniques, which can make a specific modification to a living organism's DNA to create mutations or introduce new alleles or new genes, advanced in the 2000s. These techniques—such as meganucleases, zinc finger nucleases, transcription activator-like effector nucleases, multiplex automated genome engineering, and clustered regularly interspaced short palindromic repeats—also may obviate the need for vector organisms. The types of organisms that can be manipulated and the types of manipulations that can be made have increased considerably with these newer techniques. This increase has, in due course, expanded the types and number of products that could be developed through biotechnology. For its purposes, the committee defined *biotechnology products* as products developed through genetic engineering or genome engineering (including products where the engineered DNA molecule is itself the "product," as in an engineered molecule used as a DNA information-storage medium) or the targeted or in vitro manipulation of genetic information of organisms, including plants, animals, and microbes. The term also covers some products produced by such plants, animals, microbes, and cell-free systems or products derived from all of the above.

Recognizing that "[a]dvances in science and technology . . . have dramatically altered the biotechnology landscape" since the Coordinated Framework was last updated in 1992 and that "[s]uch advances can enable the development of products that were not previously possible," the Executive Office of the President (EOP) issued a memorandum to EPA, FDA, and USDA on July 2, 2015 (EOP, 2015:2). That memorandum contained three directives to the regulatory agencies. First, they were to develop an update to the Coordinated Framework to clarify their roles and responsibilities with regards to regulating products of biotechnology. Second, as parties to the Emerging Technologies Interagency Policy Coordination Committee's Biotechnology Working Group (hereafter referred to as the Biotechnology Working Group),[1] they were to formulate (EOP, 2015:3)

> a long-term strategy to ensure that the Federal regulatory system is equipped to efficiently assess the risks, if any, associated with future products of biotechnology while supporting innovation, protecting health and the environment, promoting public confidence in the regulatory process, increasing transparency and predictability, and reducing unnecessary costs and burdens.

Third, they were to (EOP, 2015:5)

> commission an external, independent analysis of the future landscape of biotechnology products that will identify (1) potential new risks and frameworks for risk assessment and (2) areas in which the risks or lack of risks relating to the products of biotechnology are well understood. The review will help inform future policy making.

EOP published an update to the Coordinated Framework (EOP, 2017) in January 2017, which was preceded by a draft version of the update and a *National Strategy for Modernizing the Regulatory System for Biotechnology Products* (EOP, 2016) in September 2016. These two documents responded to the first two directives of the July 2015 memorandum. The present report by the National Academies of Sciences, Engineering, and Medicine responds to the memorandum's third directive.

---

[1]The Emerging Technologies Interagency Policy Coordination Committee's Biotechnology Working Group included representatives from EOP, FDA, EPA, and USDA.

BOX 1-1
Statement of Task

An ad hoc committee of the National Academies of Sciences, Engineering, and Medicine will produce a report designed to answer the questions "What will the likely future products of biotechnology be over the next 5–10 years? What scientific capabilities, tools, and/or expertise may be needed by the regulatory agencies to ensure they make efficient and sound evaluations of the likely future products of biotechnology?"

The committee will

- Describe the major advances and the potential new types of biotechnology products likely to emerge over the next 5–10 years.
- Describe the existing risk-analysis system for biotechnology products including, but perhaps not limited to, risk analyses developed and used by EPA, USDA, and FDA, and describe each agency's authorities as they pertain to the products of biotechnology.
- Determine whether potential future products could pose different types of risks relative to existing products and organisms. Where appropriate, identify areas in which the risks or lack of risks relating to the products of biotechnology are well understood.
- Indicate what scientific capabilities, tools, and expertise may be useful to the regulatory agencies to support oversight of potential future products of biotechnology.

Human drugs and medical devices will not be included in the purview of the study per a sponsor's request.

THE COMMITTEE AND ITS CHARGE

At the request of the regulatory agencies, the National Academies of Sciences, Engineering, and Medicine (hereafter referred to as the National Academies) convened a committee of experts to conduct the study "Future Biotechnology Products and Opportunities to Enhance Capabilities of the Biotechnology Regulatory System." Committee members were selected because of the relevance of their experience and knowledge to the study's specific statement of task (Box 1-1), and their appointments were approved by the President of the National Academy of Sciences in early 2016. Committee members for National Academies studies are chosen for their individual expertise, not their affiliation to any institution, and they volunteer their time to serve on a study. The present committee comprised experts with backgrounds in diverse disciplines, including biotechnology regulatory law, agricultural and industrial biotechnology, risk assessment, social science, biochemistry, engineering, entomology, microbiology, and environmental toxicology. Biographies of the committee members are in Appendix A. The study was sponsored by EPA, FDA, and USDA.[2]

INFORMATION-GATHERING ACTIVITIES

National Academies committees often invite speakers to make presentations in order to gather information relevant to the study's statement of task. The Committee on Future Biotechnology Products and Opportunities to Enhance Capabilities of the Biotechnology Regulatory System heard from 74 speakers over the course of three in-person meetings and eight webinars. All meetings and webinars were open to the public, streamed over the Internet, and recorded and posted to the study's

---

[2]The study was supported by a contract with EPA. Contract funding was provided by EPA, FDA, and USDA.

website.[3] The agendas for the meetings, topics for the webinars, and names of invited speakers can be found in Appendix B.

The committee also submitted a request for information about research on future biotechnology products and regulatory science to 28 different federal offices (see Appendix C). It received responses from 17 offices, 12 of which had publicly available information relevant to the committee's request. The submitted written responses with information can be found in the study's public access file.[4]

As with all National Academies studies, members of the public were welcome to attend meetings in person or to watch them over the Internet. At the three in-person meetings, there were opportunities for members of the public to make statements to the committee. Written comments could also be submitted to the committee at any time during the study process.[5] The comments were reviewed by the committee and are also archived in the study's public access file.

Additionally, the committee reviewed several National Academies reports that addressed aspects of future biotechnology products. In particular, it looked at *Industrialization of Biology: A Roadmap to Accelerate the Advanced Manufacturing of Chemicals* (NRC, 2015), *Gene Drives on the Horizon: Advancing Science, Navigating Uncertainty, and Aligning Research with Public Values* (NASEM, 2016a), and *Genetically Engineered Crops: Experiences and Prospects* (NASEM, 2016b) for information about forecasts of future products derived from advances in biotechnology techniques and forecasts of any potential risks associated with these products. The committee gathered information from a broad range of references that included peer-reviewed scientific literature, as well as relevant reports from agency websites and news outlets, among a variety of other sources as indicated.

ORGANIZATION OF THE REPORT

The next chapter begins with a general overview of the technical, economic, and social drivers that were influencing the types of biotechnology products being developed at the time the committee was writing its report. Chapter 2 also describes the classes of future products that the committee identified as particularly new and potentially challenging to the regulatory agencies. Examples of likely future products are given.

Chapter 3 reviews the roles and authorities of the different regulatory agencies that participate in the Coordinated Framework and describes the risk analyses the agencies used for biotechnology products. Chapter 4 analyzes whether the future biotechnology products that the committee saw on the horizon will pose different types of risks as compared to existing products and organisms. It also assesses what scientific capabilities, tools, and expertise the regulatory agencies may need to oversee these products relative to the current state of their scientific capabilities, tools, and expertise. Chapter 5 describes opportunities to enhance the capabilities of the biotechnology regulatory

---

[3]Recordings of the presentations made to the committee at its meetings and webinars can be found at http://www.nas.edu/biotech.

[4]The committee received written responses from the Army Research Laboratory Institute for Collaborative Biotechnologies, the National Science Foundation's Division of Chemical, Bioengineering, Environmental, and Transport Systems, the National Science Foundation's Division of Industrial Innovation & Partnerships, the National Science Foundation's Division of Social and Economic Sciences, the U.S. Army Corps of Engineers, the U.S. Department of the Interior's National Invasive Species Council Secretariat, the Office of Naval Research, the Defense Threat Reduction Agency, the U.S. Department of Energy's Office of Biological and Environmental Research, the U.S. Environmental Protection Agency, and the U.S. Department of Agriculture. The Intelligence Advanced Research Projects Agency responded to the request for information in its webinar presentation to the committee on July 25, 2016. Requests for the public access file can be directed to the National Academies' Public Access Records Office at PARO@nas.edu.

[5]For more information about the National Academies study process, see http://www.nationalacademies.org/studyprocess.

system, and the report concludes in Chapter 6 with the committee's primary conclusions and recommendations, which are based on its review presented in the preceding chapters.

REPORT CONTEXT AND SCOPE

The committee's work focused on future products of biotechnology and opportunities to enhance the capabilities of the regulatory system. With the diversity and number of future products anticipated over the next 5–10 years, the committee viewed the regulatory agencies, with their existing regulatory authorities granted under relevant statutes, as part of a large collection of interdependent parties involved in biotechnology-product discovery and development, science-based risk evaluation for potential entry into the marketplace, and oversight of biotechnology-product use. Figure 1-1 provides a high-level view of the many activities that can be a part of the overall regulatory landscape, including the different communities of interest and the relationship between the activities.

In the view of the committee, there are three primary phases of regulatory oversight:

- Premarket: Product development, prior to submission of a product for regulatory assessment or entry into the marketplace.
- Submitted/Premarket: Risk analysis for those products that require premarket assessments.
- Post-market: Product is available to consumers.

These three phases include multiple points of interaction between the regulatory agencies and technology developers, product developers, product consumers, and society as a whole.

For simplicity, the committee focused on three primary categories of actors or participants: *product developers*, whether these are traditional companies, small- and medium-sized enterprises, the do-it-yourself biology (DIYbio) community, or even individuals; *regulatory agencies,* including EPA, FDA, and USDA, but also other relevant agencies such as the Consumer Product Safety Commission and the Occupational Safety and Health Administration; and representatives of *society and the public*, including universities and other research organizations who might develop new biotechnology techniques, nongovernmental organizations that may be supporting or opposing specific biotechnology techniques and products, and international organizations, governments, and treaty authorities. There may be overlap between these categories, for example when a university or government laboratory is developing a technology that could itself be considered a product, either through transition of that technology to a startup or other company or through the direct use of that technology by the DIYbio community.

The interactions between the different participants in the biotechnology regulatory system vary based on the type of product, the phase of development or deployment, and whether or not a product falls under one or more statutes. For new products of biotechnology—that is, types of products whose uses have not previously been addressed by regulatory agencies—these interactions would usually be iterative in nature, with successive rounds of technology development, product development, product testing, public participation or oversight, and premarket or post-market oversight. The arrows between the different activities indicate possible interactions between parties and the types of information that are being shared—for example, submission of regulatory materials or decisions, informal communications at meetings and conferences, and formal communications through databases or public media.

The activities shown in Figure 1-1 will vary for different types of products and by the authorities and associated regulations of each agency. For example, FDA divides its product designations into premarket approval (which includes new drugs, Class III premarket-approval medical devices, and food additives), premarket notification (for products such as new dietary ingredients and

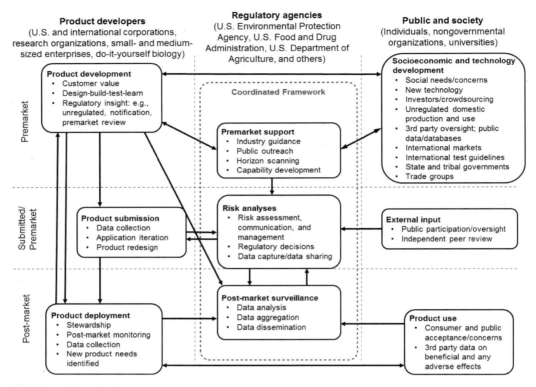

FIGURE 1-1 Schematic diagram of the regulatory system showing the primary groups involved with development, oversight, and use of biotechnology products (horizontal axis), the different phases in a product's life cycle (vertical axis), and the interactions between the various entities and activities (arrows).

NOTES: The phase on the left vertical axis indicates the life-cycle phases of biotechnology products that are intended to be sold, distributed, or marketed across state lines. The top and bottom horizontal arrows reflect the interactions among entities developing and marketing future products and users or consumers of these products as well as interactions with entities developing new technology that may support development of future products. The three vertical columns and associated arrows reflect opportunities for interaction among developers, consumers, and interested and affected parties and regulatory agencies throughout the life cycle of a product. Note that the upper-right box also addresses domestic production and use of biotechnology products that are not intended to be sold, distributed, or marketed and therefore may fall outside existing statutes, with some exceptions (described in Chapter 3). Research organizations (including universities and government research laboratories) can serve as both product developers and members of the public and society, depending on the phase and type of their activities. The arrows connecting the various activities represent some of the primary interactions between the underlying actors and processes. Note that these arrows describe potential interactions, some of which may not be present in the current regulatory framework.

Class II Section 510(k) medical devices), post-market notification (applicable to structure or function claims for dietary supplements), post-market surveillance (for products generally recognized as safe and prior-sanctioned food ingredients, pre-1994 dietary ingredients, Class I medical devices, and cosmetic ingredients), and compliance with FDA standards (for example, pre-1972 nonprescription monograph drugs). For each of these categories, all of the FDA enforcement authorities (that is, seizure, injunction, criminal prosecution, warning letters, and publicity) apply once a product is marketed. The broad set of phases used in Figure 1-1 and the categorizations used below are intended as a generalization that captures the different types of activities that might be present, independent of a specific agency's statutory structure, and help highlight some of the challenges that will be faced in oversight of future products of biotechnology.

Product Premarket Evaluation

To illustrate the flow of information in the premarket phase of the regulatory system, one can consider iterations of technology development in which university researchers present information in papers or at conferences, while product developers try out those ideas in their laboratories (arrow between "Product development" box and "Socioeconomic and technology development" box, Figure 1-2, arrow A). The regulatory agencies may be involved as participants in scientific conferences (arrow between "Premarket support" box and "Socioeconomic and technology development" box, Figure 1-2, arrow B). The developer may initiate preliminary discussions with a regulatory agency and sharing of preliminary data to obtain advice about regulatory paths (arrow between "Product development" and "Premarket support" boxes, Figure 1-2, arrow C). The use of the "design-build-test-learn" cycle of product development may occur multiple times prior to submission of a formal application for use or experimental testing. In some cases, the developer may decide to return to the ideation phase or incorporate new technology that has appeared since the initial product development.

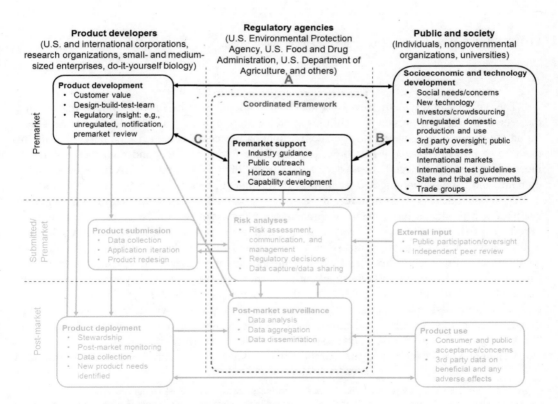

FIGURE 1-2 Premarket phase of product development.

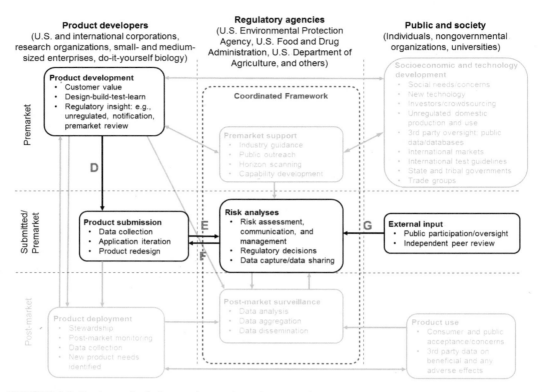

FIGURE 1-3 Product submission, review, and regulatory testing.

These discussions with the developer may lead to internal activities at the agencies that identify the need to develop insights and capabilities in preparation for anticipated regulatory decision making in the future. The agencies may decide multiple parties could benefit from a horizon-scanning exercise to inform *risk-analysis* approaches[6] for these future products and hold public outreach meetings to gain insights and knowledge (arrows from "Premarket support" box to "Product development" and "Socioeconomic and technology development" boxes, Figure 1-2, arrows B and C).

As product development matures, and assuming that the relevant statute requires a premarket evaluation, a product's more detailed "premarket" information would be prepared (arrow from "Product development" box to "Product submission" box, Figure 1-3, arrow D) and submitted to the appropriate regulatory agency or agencies (arrow from "Product submission" box to "Risk analyses" box, Figure 1-3, arrow E). Submissions may involve requests for permission to do field trials to produce data in support of an application to enter the marketplace. Depending on the results of initial studies to support the risk analysis, the regulatory agency and the product developer may determine there is sufficient certainty in the risk analysis to proceed with a regulatory decision or they may conclude that additional information may be needed for the risk analysis (arrow from "Risk analyses" box to "Product submission" box, Figure 1-3, arrow F).

Depending on the familiarity and complexity of the new product, the agency may determine in

---

[6]According to the Society for Risk Analysis, *risk analysis* includes risk assessment, risk communication, risk management, and policy relating to risk to human health and the environment, in the context of risks of concern to individuals, to public, private, and nongovernmental organizations, and to society at a local, regional, national, or global level.

some instances that public participation concerning the technology, the benefits of the technology, and its potential implications would be helpful to inform the risk analyses. The agencies may also engage external groups for peer review (possibly via a federal advisory committee) of a draft risk analysis (arrow from "External input" box to "Risk analyses" box, Figure 1-3, arrow G). For new product submissions that are less complex or are more familiar to the regulatory agency or agencies, these steps could be less involved or omitted.

Assuming the regulatory decision supports entry into the marketplace, the developer can deploy the product consistent with the terms of the regulatory decision (arrow from "Risk analyses" box to "Product submission" box and arrow from "Product submission" box to "Product deployment" box, Figure 1-4, arrows H). At this stage, the product is available to consumers or users. Depending on the operative statute, if the developer wishes to expand the use pattern of the product, it may be required to submit additional data to the appropriate regulatory agency (or agencies) for evaluation. Depending on the agency and the results of the risk analysis, the developer may be required to generate post-market data to refine the existing risk analysis (arrow from "Product deployment" box to "Post-market surveillance" box, Figure 1-4, arrow I). Insights from consumers and users of the product, including data on product performance or unanticipated adverse effects (arrow from "Product use" box to "Post-market surveillance" box, Figure 1-4, arrow J), could also prompt an agency to reevaluate the existing risk analysis. Interactions between the product developer and users or other third parties could also lead to new product development within a company, new research

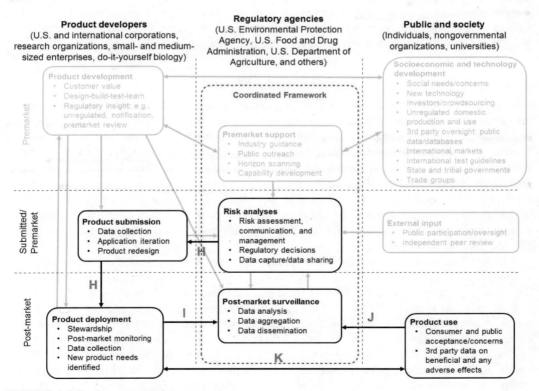

FIGURE 1-4 Product post-market assessment.

funded by government or other sources, or new discussions of social needs, benefits, and concerns (arrow between "Product deployment" and "Product use" boxes, Figure 1-4, arrow K).

Product Notification Alternative

A slightly different scenario is one in which no premarket approval is required. In this scenario, the initial premarket discussions described above could still occur, but in the context of premarket support it could be determined that, given the relevant statute for the product, a premarket evaluation is not required. In this case, the developer may only need to notify the appropriate agency that the product is entering the market and the developer has determined the product meets the relevant statute's safety standard. In this scenario, if post-market data indicate unanticipated adverse effects associated with the product's use, the appropriate regulatory agency could initiate a risk analysis to determine if risk mitigation is required.

Unregulated Products Alternative

In this scenario, initial premarket discussions with a product developer may determine that, given the nature of the product, it is not regulated under existing statutes. In this instance, the developer may voluntarily enter discussions with the relevant agency (or agencies) prior to entering the marketplace and subsequently rely on its stewardship program to determine if any changes to the product's patterns of use are needed. There could also be instances where a product is being manufactured within a home for domestic use, with no intention to market, distribute, or sell the product over state lines. Depending on the nature of the product and the relevant statute, the product may not be subject to regulation, even though if it were manufactured by a company it could be subject to regulatory oversight. In this instance, the regulatory agencies could partner with third-party organizations to help support the development of stewardship programs.

Other Interactions

As noted in Figure 1-2 (arrow B), the regulatory agencies could interact with other parties on issues related to implementing the Coordinated Framework. For example, the agencies could work with various global organizations (such as the Organisation for Economic Co-operation and Development) to develop international test guidelines for future biotechnology products. These collaborative efforts could expand technical capability and enhance efficiency in developing risk-analysis methods and also enhance efficiency and effectiveness in evaluating products intended for U.S. import. Harmonized guidelines also support efficiencies for U.S. developers intending to export their products. Finally, the agencies may wish to interact with third parties to advance research and development of new testing techniques, risk models, and other techniques to support risk analyses for future products.

In developing the recommendations of its report, the committee assessed the opportunities for enhancing the capabilities of the U.S. biotechnology regulatory system through interactions with this broad community of interested and affected parties. These interactions are especially important in areas of rapid technology change, such as biotechnology, where the regulatory agencies must maintain adequate capability to allow appropriate assessments of new technologies that go beyond existing biotechnology products and for which there may not yet be well-established approaches to risk analyses. These interactions are especially important in areas where there may be substantial public discussion regarding the risk–benefit tradeoffs of a technology, ethical considerations, and competing international activities that influence U.S. policy and trade.

The committee did not consider it to be part of its task to comment on the structure of the U.S.

regulatory system and whether it was optimally situated to provide appropriate oversight of future biotechnology products. Rather, it focused on the current system as described in the update to the Coordinated Framework (EOP, 2017) and tried to articulate the extent to which future products of biotechnology would generate new types of risks and to identify the opportunities for enhancing the capabilities and capacity of the biotechnology regulatory system to handle types of products that the committee saw on the horizon.

REFERENCES

Cohen, S.N., A.C.Y. Chang, H. Boyer, and R.B. Helling. 1973. Construction of biologically functional bacterial plasmids in vitro. Proceedings of the National Academy of Sciences of the United States of America 70:3240–3244.

EOP (Executive Office of the President). 2015. Memorandum for Heads of Food and Drug Administration, Environmental Protection Agency and Department of Agriculture. July 2. Available at https://obamawhitehouse.archives.gov/sites/default/files/microsites/ostp/modernizing_the_reg_system_for_biotech_products_memo_final.pdf. Accessed January 31, 2017.

EOP. 2016. National Strategy for Modernizing the Regulatory System for Biotechnology Products. Available at https://obamawhitehouse.archives.gov/sites/default/files/microsites/ostp/biotech_national_strategy_final.pdf. Accessed January 31, 2017.

EOP. 2017. Modernizing the Regulatory System for Biotechnology Products: An Update to the Coordinated Framework for the Regulation of Biotechnology. Available at https://obamawhitehouse.archives.gov/sites/default/files/microsites/ostp/2017_coordinated_framework_update.pdf. Accessed January 30, 2017.

NASEM (National Academies of Sciences, Engineering, and Medicine). 2016a. Gene Drives on the Horizon: Advancing Science, Navigating Uncertainty, and Aligning Research with Public Values. Washington, DC: The National Academies Press.

NASEM. 2016b. Genetically Engineered Crops: Experiences and Prospects. Washington, DC: The National Academies Press.

NRC (National Research Council). 2015. Industrialization of Biology: A Roadmap to Accelerate the Advanced Manufacturing of Chemicals. Washington, DC: The National Academies Press.

OSTP (Office of Science and Technology Policy). 1986. Coordinated Framework for Regulation of Biotechnology. Executive Office of the President. Federal Register 51:23302. Available at https://www.aphis.usda.gov/brs/fedregister/coordinated_framework.pdf. Accessed September 24, 2016.

2

Emerging Trends and Products of Biotechnology

This chapter describes the technical, economic, and social trends that will drive the development of biotechnology products likely to emerge over the next 5–10 years. Advances in biotechnology, new actors, economic investments, and societal challenges and concerns all influence the new types of biotechnology products in development. The chapter also outlines the changes in the scope, scale, complexity, and tempo of biotechnology products, which the committee believes will lead to a profusion in the next decade of products made through the use of biotechnology. Types of products likely to be developed—and the kinds of environments in which they may be used—are reviewed.

SETTING THE STAGE: UNDERSTANDING THE KEY DRIVERS FOR FUTURE BIOTECHNOLOGY PRODUCTS

Increasing investment in the bioeconomy, complex societal challenges, the confluence of new technical drivers, and a proliferation of new actors are transforming both biotechnology products and the context in which the U.S. regulatory system operates. For this reason, it is important to track changes in multiple areas that may affect product development and penetration rates. To help set the stage about who and what is influencing the development of new biotechnology products, this section gives a brief overview of a number of these drivers and some of their possible effects on regulation of future products of biotechnology.

Technical Drivers

Several technical drivers have increased the rate at which new products can be developed and also increased the accessibility of modern tools of biotechnology, resulting in an increased number of actors who are able to create biotechnology products. Some key areas include DNA sequencing, synthesis, and editing; standardization of biological parts; and increasingly rapid design-build-test-learn cycles.

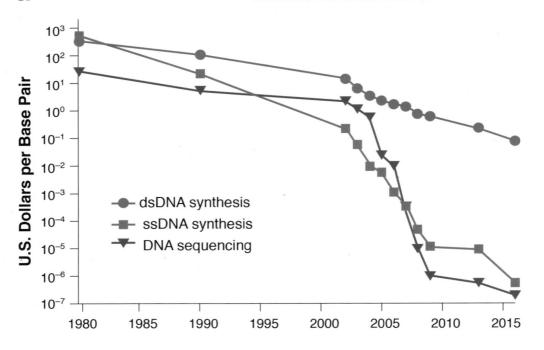

FIGURE 2-1 Efficiency trends in DNA sequencing (green) and synthesis of double-stranded DNA (dsDNA, blue) and single-stranded DNA (ssDNA, red), 2001–2015.
SOURCE: Adapted from Boeke et al. (2016).

New Tools for DNA Sequencing, Synthesis, and Editing

The cost of sequencing DNA dropped by seven orders of magnitude between 2002 and 2008 and has dropped by an additional order of magnitude between 2008 and 2015 (Figure 2-1), representing a rate of decrease that substantially exceeded that of Moore's law in the 2002–2008 time frame. The price of double-stranded DNA synthesis has also decreased exponentially fast, though at a lower rate, which has enabled the creation of more complex synthetic constructs (Zhou et al., 2015; Chambers et al., 2016), including the construction of whole genomes (Juhas and Ajioka, 2017). In addition, the emergence of genome-engineering technologies (Box 2-1) has enabled targeted modification of DNA sequences—such as insertions, deletions, and site-specific replacements of DNA bases—in a variety of organisms, affording a host of applications (Gaj et al., 2013; Reardon, 2016). Similarly, advances in the understanding of how RNA interference (RNAi) silences gene expression are creating opportunities to create new products, for example for insect-pest control in agricultural crops. Taken together, these trends have made it possible to "mine" genetic data from a wide variety of organisms and then to synthesize new genetic constructs that modify the function of living organisms.

Standardized Biological Parts

The field of synthetic biology has promoted the use of open-access, standardized parts for over a decade. The Registry of Standard Biological Parts[1] was established at the Massachusetts Institute

---

[1]Registry of Standard Biological Parts. Available at http://parts.igem.org/Main_Page. Accessed September 27, 2016.

BOX 2-1
The Evolution of Gene-Scale to Genome-Scale Engineering Technologies

The tools used to create biotechnology products have evolved dramatically since recombinant-DNA technology was developed in the 1970s. That technology has given way to genome-engineering technologies that have been fueled by *de novo* DNA synthesis, large-scale gene assembly, and genome editing. Such technologies increase the rate at which products can be developed because of the scale, speed, and precision at which desirable changes can be made in the DNA of an organism.

Recombinant DNA (rDNA) techniques join together DNA molecules that are not found together in nature and then insert these joined molecules into a host organism to create new genetic combinations (Cohen et al., 1973). Recombinant DNA insertions typically involve the transformation of plasmids—the insertion of circular rDNA molecules capable of replicating alongside the host's genome—or direct insertion into the host's genome using either the natural ability of cells to exchange DNA or by techniques that create double-stranded breaks in the genome, allowing incorporation of exogenous DNA. The understanding of how to harness rDNA opened up the door to genetic engineering. In the 1970s, scientists learned that one way to deliver DNA into plant cells was through the use of *Agrobacterium tumefaciens*, a bacterium that transfers tumor-inducing plasmids into the cells of host organisms in nature. The plasmids could be engineered to eliminate undesirable tumor-inducing genes and encoded with genes with desirable traits that would be taken up when the transformed *Agrobacterium* are inserted in a host cell. Another gene delivery method is to coat metal particles with DNA and pierce target cells with these particles, using what is referred to as microprojectile bombardment, gene gun, or biolistics (Klein et al., 1987). When the DNA of a source organism is delivered into a host cell, a natural recombination event can happen, where the source DNA replaces the host DNA. A challenge with rDNA is that the location where the source DNA recombines with the host DNA to replace it can be random. It is a time-consuming process to deliver source DNA into many cells and then identify the cells where source DNA was inserted in a useful place. Millions of cells must be screened to identify those that are worth growing into genetically engineered microorganisms, plants, and animals.

The advent of genome engineering—which uses tools that allow rapid and precise changes directly across chromosomes of living cells instead of limited modifications at single genes using rDNA methods—has transformed basic and applied biological research. The confluence of three complementary technologies has fueled advances in genome engineering. First, *de novo* DNA synthesis technologies enable tailored construction of user-defined double-stranded DNA segments. Similar to advances in DNA sequencing, DNA synthesis has undergone logarithmic improvements in scale, cost, and throughput. For example, large-scale DNA microchip-based synthesis methods permit high-density synthesis of around 10^5 customized single-stranded DNA (oligonucleotides). Gene synthesis—the overlap of these oligonucleotides—combined with improvements in DNA error-correction methods has enabled high-quality and cheap construction of designer synthetic genes (Kosuri and Church, 2014). Second and complementary to DNA synthesis, large-scale DNA assembly in vivo and in vitro methods permit the precise assembly of individual synthetic genes into higher-order combinations at the network and whole-genome scales. Advanced recombination and transplantation techniques are in development to improve the efficiency of introduced synthesized genes, pathways, and genomes into target organisms (Gibson et al., 2010). Finally, *genome editing* (often used interchangeably with the term *gene editing*) permits targeted changes directly in the chromosomes of living cells. Many genome-editing technologies generate DNA double-strand breaks at targeted loci to introduce genomic modifications. There are four main classes of sequence-specific nucleases (reviewed by Voytas and Gao, 2014): meganucleases, zinc finger nucleases (ZFNs), transcription activator-like effector nucleases (TALENs), and the clustered regularly interspaced short palindromic repeat (CRISPR)-Cas9 nuclease system. Meganucleases were discovered first, followed by ZFNs and TALENs. Although ZFNs and TALENs recognize specific DNA sequences through protein–DNA interactions and use the *FokI* nuclease domain to introduce double-strand breaks at genomic loci, construction of functional ZFNs and TALENs with desired DNA specificity remains laborious, costly, and primarily limited to modifications at a single genetic locus. CRISPR-Cas9 has been broadly adopted for multiplexed targeting of genomic modifications because the CRISPR nuclease Cas9 uses a short guide RNA to recognize the target DNA via Watson-Crick base pairing and has been shown to function in many organisms (Cong et al., 2013; Jinek et al., 2013; Mali et al., 2013). The CRISPR-Cas9 nuclease system is an innate bacterial defense mechanism against viruses and plasmids that uses RNA-guided nucleases to

continued

BOX 2-1 Continued

cut the foreign DNA sequences and thereby disable them. Scientists have reengineered the CRISPR-Cas9 system so that a single RNA (the guide RNA) can direct the Cas9-mediated cut of a target sequence in a genome. The ease of design and the specificity and simplicity of the CRISPR-Cas9 system have made it a popular technique for generating future biotechnology products (Doudna and Charpentier, 2014). Nevertheless, other genome-editing approaches are still emerging. For example, multiplex automated genome engineering (MAGE) permits multisite genome modifications through hybridization of synthetic oligonucleotides during the process of DNA replication. MAGE enacts editing at base-pair precision at high efficiencies and has been used for pathway diversification (Wang et al., 2009), whole-genome recoding (Isaacs et al., 2011; Lajoie et al., 2013), and molecular evolution of proteins (Amiram et al., 2015). At the time the committee's report was being written, non-Cas9 nucleases had been recently described for CRISPR genome editing (Zetsche et al., 2015).

Leveraging these advances in genome engineering, *synthetic biology* has also been used to generate new products. In synthetic biology, engineering principles are applied to reduce genetics into DNA "parts" so that those parts can be understood in isolation and reassembled into new biological parts, devices, and whole systems to build desired functions in living cells. Through this process, it is possible to assemble new organisms from parts of DNA from more than one source organism or to build synthetic DNA from molecules.

---

of Technology as part of the international Genetically Engineered Machine (iGEM) competition, a student synthetic-biology research competition started in 2004. In 2009, the BIOFAB: International Open Facility Advancing Biotechnology (the first biological design-build facility) placed more than 2,500 standardized, quantitatively defined, biological parts in the public domain. The biological parts are compatible, minimal DNA sequences that code for distinct biological functions, such as coding sequences that are responsible for the expression of proteins, promoters, and terminators (that cause transcription to cease). There is also a repository for scientists to share plasmids.[2] In 2016, the National Institute of Standards and Technology partnered with Stanford University on a project in Silicon Valley focused on facilitating the standardization of parts relevant to biology.[3]

All of these efforts are intended to leverage the types of advances that occurred in mechanized production or the later information-technology revolution that relied on standardized parts with reproducible characteristics, combined with increasingly accurate measurement tools. Historically, advances in standardization have been linked to more rapid innovation, improved in-company research and development efforts, increased length and complexity of supply chains, better efficiency and quality management, and greater network effects (including trade). The use of standard parts has accelerated the assembly of complex products in other fields, such as the automotive industry, and such advances may lie ahead for biotechnology as well. Engineering of biological function will increasingly need accurate documentation of components, including description and performance characteristics, combined with incentives to make parts more modular and more predictable, with the critical goal of driving out defects in the field.

The potential importance of standardized biological parts, or other components of biotechnology, is multifold. It enables the reuse of previously engineered devices, creating the ability to design more complex systems more predictably, more rapidly, and with fewer failures. It also enables a

---

[2]Addgene. Available at https://www.addgene.org. Accessed February 13, 2017.

[3]The project is called the Joint Initiative for Metrology in Biology. Available at http://jimb.stanford.edu. Accessed September 27, 2016. The National Institute of Standards and Technology is part of the U.S. Department of Commerce.

wider variety of practitioners to make use of advances of biotechnology by packaging advances in a form that can be reused and matured by others, and it is a first step toward a biotechnology development "ecosystem" in which different companies specialize in components and subsystems, allowing others to make use of the advances across a wide variety of areas at an increased pace.

The use of standardized parts also provides an opportunity for improving regulatory analysis by incorporating safety features that enhance regulatory assessment (similar, perhaps, to the safety ratings issued by the company UL). Achieving safety in complex systems is typically not a simple process and the enormous complexity and variety of biological organisms will be challenging in this regard. Standardization of components provides a possible means to enable more rigorous safety standards and protocols for safety certification, of the sort that is seen in the automotive and civilian aerospace industry.

Increase in the Speed of the Design-Build-Test-Learn Cycle

Traditionally, biotechnology has been challenged by reproducibility issues—an engineered microbe might stop producing or the production rate could fluctuate. Scaling production from micrograms to kilograms and potentially to kilotons was an expensive, high-risk, and costly process. Predictive modeling and computer-aided design tools common in other engineering disciplines were almost nonexistent in biology.

Sun et al. (2014) noted that "[d]ecreasing the design-build-test cycle length is a fundamental challenge facing all engineering disciplines. This is acutely true in synthetic biology." The design-build-test-learn (DBTL) cycle (Figure 2-2) is the "fundamental building block of effective and

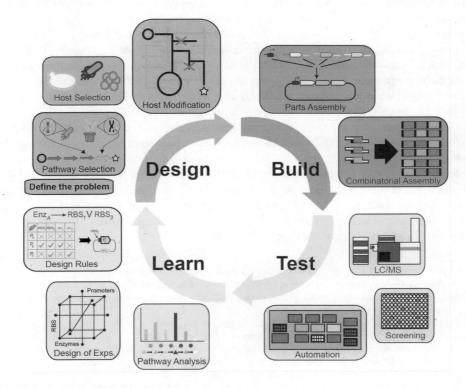

FIGURE 2-2 Iterative design-build-test-learn process.
SOURCE: Petzold et al. (2015).

efficient problem solving" (Wheelwright and Clark, 1994:34). It has been adapted for the engineering of biological systems. The design component defines the problem, establishes an approach to solve the problem, and identifies the biological components needed to build or modify. The build component synthesizes, assembles, or edits (or all three) the components of the engineered biological system. The test component characterizes the different biological systems and identifies the variants with the prescribed behavior. The learn component analyzes the test data and informs subsequent iterations of the cycle. The ability to consistently shorten the DBTL cycle can reduce lead times for product development and translate to competitive advantages, especially for those firms who enter markets early with superior products and can command premium prices (Gibson et al., 2010; Gill et al., 2016; Hutchison et al., 2016). However, increases in the efficiency and speed of DBTL cycles also provide challenges and opportunities in the area of safety, where the use of well-established engineering workflows needs to be optimized to ensure that both intermediate and final designs can be developed and deployed in a safe manner.

Additional Technical Drivers

Although advances in DNA sequencing, synthesis and editing, standardization of biological parts, and increasingly rapid DBTL cycles are fundamental technical drivers, there are many other interrelated drivers that play important roles. The expansion of public and private biofoundries—centralized facilities that leverage software and automation to dramatically increase the number of organisms that can be engineered in parallel (Eisenstein, 2016)—is expected to have a substantial effect on the rate of introduction of biotechnology products to the marketplace, perhaps enabling academic laboratories and companies to complete development of multiple biotechnology products per year. The existence of "open-source" approaches to synthetic biology, such as the iGEM parts registry, are enablers for high schools, universities, and the do-it-yourself biology (DIYbio) community to develop courses that teach students and interested parties about engineering biology. Digitization, including not only the representation of DNA in digital repositories but the increased use of standard computer markup languages—such as SBML[4] and SBOL[5]—enable sharing of information about biological models and designs. These are in turn enablers for better tools for predictive models that can be used in the design process, such as SimBiology,[6] Clotho,[7] and TinkerCell.[8]

Additional drivers that are not specific to future products of biotechnology but are nonetheless enablers of increasingly rapid product innovation are peer-to-peer sharing platforms such as Benchling,[9] which provides software tools for software solution for experiment design, note taking, and molecular biology, or OpenWetWare,[10] which provides an open wiki for synthetic biologists; "cloud-based" experimental platforms such as Transcriptic[11] and Emerald Cloud Lab[12] that provide access to advanced instrumentation and automation on a fee-for-service basis; and a variety of biotechnology incubator spaces, such as QB3[13] and LabCentral,[14] that enable biotechnology startups to have access to advanced laboratory facilities.

---

[4]Systems Biology Markup Language. Available at http://sbml.org. Accessed October 11, 2016.

[5]Synthetic Biology Open Language. Available at http://sbolstandard.org. Accessed October 11, 2016.

[6]Simbiology: Model, simulate, analyze biological systems. Available at http://www.mathworks.com/products/simbiology. Accessed October 11, 2016.

[7]Clotho. Available at https://www.clothocad.org. Accessed October 11, 2016.

[8]TinkerCell. Available at http://www.tinkercell.com. Accessed October 11, 2016.

[9]Benchling. Available at https://benchling.com. Accessed October 11, 2016.

[10]OpenWetWare. Available at http://openwetware.org. Accessed October 11, 2016.

[11]Transcriptic. Available at https://www.transcriptic.com. Accessed October 11, 2016.

[12]Emerald Cloud Lab. Available at http://emeraldcloudlab.com. Accessed October 11, 2016.

[13]QB3. Available at http://qb3.org. Accessed October 11, 2016.

[14]Lab Central. Available at http://labcentral.org. Accessed October 11, 2016.

Impact on Regulation

The combination of technical drivers described above has increased the rate at which new biotechnology products can be created, the scope and complexity of those products, and the number and type of actors who engineer new biotechnology products. In the past, many developers of biotechnology products have been established companies that have strong knowledge of the regulatory system, but when the committee was writing its report there were an increasing number of small- and medium-sized enterprises and DIYbio enthusiasts who were developing technologies and products. Handling the increased scale of products and diversity of developers will require a regulatory system that is agile enough to rapidly adapt to technological change.

Economic Drivers

A second area of rapid change is in the economic drivers that underlie the development of new biotechnology products. Although difficult to accurately determine, total domestic revenues in 2012 from biotechnology—biological, agricultural, and industrial biotechnology products derived using genetically engineered (GE) organisms—have been estimated to be at least $324 billion and to have grown at a rate of more than 5 percent of U.S. gross domestic product annually from 2007 to 2012 (Carlson, 2016).[15] Interest from governments and the private sector contributes significantly to this growth.

Government Investment in Biotechnology Products

More than 40 countries, including the United States, have created national strategies or domestic priorities for developing and promoting a 21st-century bioeconomy (EC, 2012; Formas, 2012; OSTP, 2012; OECD, 2015; El-Chichakli et al., 2016). According to the Executive Office of the President's Office of Science and Technology Policy, the *bioeconomy* is "research and innovation in the biological sciences [used] to create economic activity and public benefit" (OSTP, 2012:7). Governmental policies for promoting the bioeconomy seek to combine technological innovation, economic growth, ecological sustainability, and resource efficiency (GBC, 2016). For example, the Obama Administration endorsed the U.S. version of this vision in its 2012 *National Bioeconomy Blueprint*, which set forth broad-based advances in biotechnology, including biobased chemicals, biofuels, and new tools to address challenges and next-generation opportunities in agriculture and manufacturing (OSTP, 2012). The rapid global growth of the bioeconomy is expected to accelerate and increase the demand for biotechnology products (Carlson, 2016).

Private-Sector Investment and Diversification of Sources of Capital

In 2015, U.S. biotechnology companies raised more than $61 billion,[16] 32 percent more than the previous record set in 2014 (EY, 2016; Figure 2-3). Roughly half of that funding went to companies with revenues of less than $500 million (EY, 2016). Venture capital investment in biotechnology companies increased more than $3 billion from 2014 to 2015, reaching $9.4 billion (EY, 2016). Although much of this money went to companies developing human drugs and medical devices, funding for nonhuman biotechnology was also strong; for example, many of the people who built the dot.com economy, from Jerry Yang of Yahoo to Eric Schmidt of Google and Peter

---

[15]This estimate includes biotechnology-based drugs and medical devices for human use, which were not part of the committee's statement of task.

[16]Biotechnology companies producing human drugs and medical devices are included in this total.

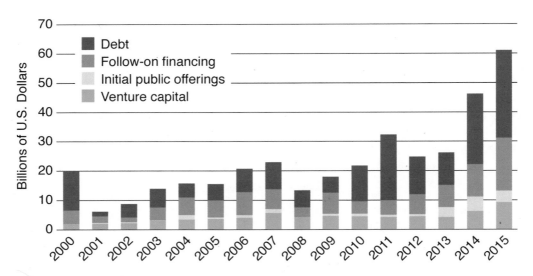

FIGURE 2-3 U.S. biotechnology financings by year.
NOTE: Biotechnology companies in the figure include those producing human drugs and medical devices, which were not within the committee's statement of task.
SOURCE: EY (2016).

Thiel of PayPal, invested heavily in synthetic-biology startup companies between 2009 and 2015 (Hayden, 2015).

Increasingly, capital for biotechnology research is not limited to governmental sources and private-sector funding. Crowdsourcing websites like Indiegogo and Kickstarter have been a source of funding for entrepreneurs and small companies for biotechnology products (see examples in Table 2-1) since they launched in 2008 and 2009, respectively, though Kickstarter announced in 2013 that it would cease funding campaigns that offered a GE organism as a reward following backlash to a campaign for a bioengineered glowing plant (Luzar, 2013). Changes in 2016 to the U.S. Securities and Exchange Commission codes affecting crowdfunding have given rise to a number of equity sites, such as WeFunder.org, which provide another option for biotech startups and entrepreneurs (SEC, 2016). Accelerator organizations like Y Combinator and IndieBio also have provided seed money for biotechnology-product developers.[17] The diversification of capital sources has played a role in the diversification of biotechnology-product developers (Box 2-2).

Impact on Regulation

The combination of new investments in biotechnology research, diversified sources of capital, and new players in the development of biotechnology products has the potential to create many new challenges for the biotechnology regulatory system. Crowdfunding and other new financing mechanisms for research and product development may place these activities outside the reach of traditional Coordinated Framework research and biosafety oversight mechanisms such as the National Institutes of Health Guidelines (see Chapter 3). New players and lower barriers to entry

---

[17]See Y Combinator: Biotech, available at https://www.ycombinator.com/biotech, and Indie Bio: Companies, available at http://sf.indiebio.co/companies-2. Accessed January 8, 2017.

TABLE 2-1 Examples of Crowdfunded Biotechnology Products

	Crowdfunding Site	Project	Year	Amount Raised
Glowing Plant	Kickstarter	Glow-in-the-dark plant	2013	$484,013
Real Vegan Cheese	Indiegogo	Milk protein from yeast	2014	$37,369
Purdue iGEM	Experiment.com	Waste-clearing *E. coli*	2016	$3,090

SOURCES: Information from Kickstarter.com, Indiegogo.com, and Experiment.com cited in Regalado (2016).

may alter the number and types of regulated entities whose activities the regulatory agencies are tasked with overseeing and may blur key jurisdictional concepts such as who is the "product sponsor," "product developer," or "manufacturer" that the agencies can regulate. These trends include DIYbio community laboratories, at-home and direct-to-consumer biotechnology developers, crowdsourced funding and idea generation, and smaller-scale and decentralized manufacturing.

Societal Drivers

In addition to technical and economic drivers, there are a large variety of societal drivers that come into play in the context of both current and future products of biotechnology. The extent to which these societal drivers are directly part of the regulatory system depends on the specifics of the agency and the authority, but they often set the stage for the discussion regarding the evaluation, oversight, and usage of a specific biotechnology product or class of products. The committee reviewed some of the background and context for these societal drivers, with a view toward the types of changes that future biotechnology products may play.

Potential Societal Benefits from Future Biotechnology Products

Biotechnology innovations may have the ability to simultaneously address societal challenges and produce economic benefits. Some biotechnology products are envisaged as tools to help address food security and climate change and to promote "green growth" and environmental sustainability. For example, a U.S. government study predicts that new biotechnology products associated with biomass production could cut U.S. greenhouse gas emissions by 400 million tons per year, or 8 percent (Biomass R&D Board, 2016). Biotechnology products that are successful at addressing societal challenges may have associated economic benefits such as increased productivity and new job creation (including jobs for higher skilled labor).

Nontraditional Players in the Development of Biotechnology

Once the purview of PhD-level researchers, biotechnology is taught in some high school science classes, and, since 2009, a growing number of DIYbio community laboratories in the United States and Europe teach basic biotechnology to nonexperts through formal classes and informal education approaches. In 2013, a survey of the DIYbio community, estimated to be between 3,000 and 4,000 people worldwide, found that the majority of the 359 respondents (82 percent) were in the United States, 10 percent were in Europe, 4 percent in Canada, 1 percent in China, and 2 percent from elsewhere. The community respondents were mostly adult males (75 percent), and few of them (less than 10 percent) work solitarily, that is, outside of community laboratory spaces where technical expertise and equipment are concentrated (Grushkin et al., 2014). Projects sup-

BOX 2-2
New Actors and Market Niches

A key, new element associated with the development of biotechnology products in the United States is the multiplicity of different players that intersect and collaborate in flexible but robust entrepreneurial ecosystems. Players includes companies of all sizes, university entrepreneurs, social entrepreneurs, venture philanthropists and foundations, the investment community, do-it-yourself biology (DIYbio) community laboratories, and citizen scientists. In 2016, there were 36 DIYbio community laboratories in the United States.[a]

As the time and costs required to experiment with biology drop—based on standardized parts, open-source libraries, digitalization, better predictive simulations, increased access to advanced molecular-biology tools, and peer-to-peer sharing platforms—actors will explore how these new capabilities can be leveraged. Increasingly, companies will tap into the innovativeness of their suppliers and customer base; consumers may use similar techniques to engage other consumers in the design and production process. This approach is already being used in a variety of sectors, from food additives to fragrances to plastics, video games, and semiconductors and is driven by a powerful objective: "test what works and what doesn't [work] as early as possible . . . and direct attention and integrated problem solving at potential downstream risks" (Thomke, 2003).[b]

Thus, the design-build-test-learn cycles are not just speeding up (Gibson et al., 2010; Hutchison et al., 2016), but the experiments are expanding to include more potential sources of innovation—a *crowdsourcing* model. It is reasonable to speculate that in 10 years (circa 2027) peer-to-peer sharing repositories will emerge that are similar to those that already store thousands of digital designs for physical objects, such as Thingiverse, Pinshape, and YouMagine.[c] Such platforms could overlap and become synergistic, supporting the creation of physical or computational tools to explore and develop biological designs and products. The rapid prototyping and development process would allow new market niches to appear: for instance, consumers selling or exchanging biotechnology tools, components, and products with other consumers or producing private solutions for themselves, family, and friends (Figure 2-4). Examples of these consumer-to-consumer approaches are Odin Labs CRISPR kit[d] and Bento Labs,[e] which sell advanced molecular-biology tools directly to consumers for use by DIYbio enthusiasts, and Peer-to-Peer Probiotics,[f] which is a startup company that encourages the development and sharing of probiotic microbial strains between consumers themselves. In these peer-to-peer digital systems, international boundaries may become increasingly irrelevant along with unilateral approaches to regulation and intellectual-property control.

ported at community laboratories involve bacteria, fungi, and plants and largely fall into educational, artistic, and commercial categories. Importantly, U.S. and European community laboratories independently develop and operate under similar codes of conduct that include shared themes of transparency, safety, open access, and education (Kuiken, 2016). DIYbio community laboratories have already developed and adopted safety protocols and provide access to biosafety professionals via a Web-based portal.[18] In addition to access to community laboratories, equipment, and biosafety professionals, the DIYbio community has increasing access to funding for their work through crowdfunding platforms, such as Kickstarter and Indiegogo, that allow public donations to support interesting projects.

The iGEM Foundation also encourages nontraditional players through its annual iGEM competition. Student teams compete during the summer to create biotechnology systems, using standardized biological parts, with the objective of making positive contributions to their communities and the world. The number of iGEM teams has grown from 5 U.S. teams in 2004 to more than 300 teams, including 40 high school teams, from 30 countries, in 2016. The distribution of iGEM teams

---

[18]Ask a biosafety professional your question. Available at http://ask.diybio.org. Accessed January 23, 2017.

	New Business to Business	Business to Consumer	Consumer to Consumer	Consumer
Structure	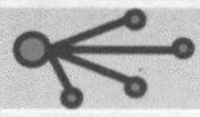	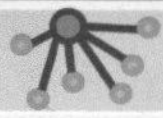		
Examples	Ginkgo Bioworks Zymergen Bolt Threads Contract synthesis	Taxa New Harvest	Shareable iGEM registry Odin Labs CRISPR kit Peer-to-Peer probiotics	At-home bioreactors Do-it-yourself products (for individuals, family, and friends)
Analogs	Vertical to horizontal Plug and play	Mass customization Customer-driven design	Artisanal products On-line sales BioEsty	Cloud-enabled, 3-D printing
Enabling Tools Drivers	Genome editing Bioinformatics 1000+ molecules	Genome editing Social media Bioinformatics 1000+ molecules	Crowdsourced designs	Cloud computing Open-source repositories Standardized parts
Governance Approaches	Existing regulations	Regulations Covenants Voluntary agreements Social-benefit corporations	Consumer preferences Codes	Codes of conduct Watermarking

FIGURE 2-4 New markets and business models for products of biotechnology.
NOTES: Products of biotechnology are being sold or exchanged in new ways as outlined by the column titles: new business to business, business to consumer, consumer to consumer, and consumer for self. For each, product or company examples are listed along with various characteristics of the products. Blue circles present businesses and orange circles represent consumers; larger circles represent the source of the product.

[a]DIY Bio, Local Groups. Available at https://diybio.org/local. Accessed September 27, 2016.
[b]In the semiconductor industry, shifting experimentation and innovation to customers helped create a custom chip market, which grew to more than $15 billion in annual revenues in over two decades.
[c]See https://www.thingiverse.com; https://pinshape.com; https://www.youmagine.com. Accessed December 3, 2016.
[d]The ODIN genetic engineering home lab kit. Available at http://www.the-odin.com/genetic-engineering-home-lab-kit. Accessed December 3, 2016.
[e]The Bento Lab. Available at https://www.bento.bio/bento-lab. Accessed December 3, 2016.
[f]Peer-to-Peer Probiotics. Available at http://p2pprobiotics.com. Accessed December 3, 2016.

from around the world differs from that of the DIYbio community survey; in 2015, the majority of iGEM teams (37 percent) were from Asia, 28 percent from North America, and 25 percent from Europe. The iGEM teams are a growing source of creative ideas and prototypes for future biotechnology products in a large number of global challenge "tracks" from a broad range of applications including therapeutics, energy, environment, food and nutrition, information processing, art and design, hardware, and software. The difference in geographical distribution between 2013 DIYbio community adult survey respondents and 2015 iGEM student teams may foreshadow an increase in future biotechnology products with origins outside the United States. A large percentage of these products could be developed with the intention of export to the United States, which would further increase the number of products U.S. regulatory agencies would have to assess in the next 5–10 years.

Researchers in universities and government laboratories are also active in researching and developing new biotechnology approaches and products, including in specialty crops where markets may be too small to appeal to big agricultural companies. For example, in the 1980s, university researchers began developing a papaya cultivar with resistance to papaya ringspot virus, a devastating disease that threatened to end papaya production in Hawaii. With the GE resistance to the virus, papaya production has continued on Hawaii; as of 2009, the resistant cultivar was planted on more than 75 percent of papaya-producing acres (USDA–NASS, 2009).

Patent expirations also provide an opportunity for nontraditional players to contribute to biotechnology advances. After the patent expired for the first commercialized soybean with GE resistance to the herbicide glyphosate, researchers at the University of Arkansas's Crop Variety Improvement Program spent a number of years mating plants with the off-patent resistance gene into a soybean that was in development at the university, essentially creating a GE generic product that offers growers a lower-priced alternative that can be planted in subsequent years without paying a technology fee that is required with patented seeds (Miller, 2014).

Societal Views of Biotechnology

Societal views about biotechnology differ widely among different regions of the world, including within the United States. Although some sectors of U.S. society see biotechnology as a way to solve the great challenges facing humanity today (for example, increase food production efficiency, reduce carbon footprint, or develop more humane farming systems), other sectors of society perceive biotechnology as having both negative and positive aspects or as a threat. The nature of the concern varies but generally pertains to categories that include physical threats to human health, the environment, biodiversity, and resource accessibility as well as other threats such as ownership of biomaterials, technological systems, agriculture or the environment, or genetic resources, and power and voice in decision making about technological choices.

The root of several of the views about biotechnology stems from differing world views about how uncertainty should be treated in decision making, what types of risks should be considered in oversight, the role of technology in addressing problems of society, and who should have power, voice, and choice. Some groups argue for the use of the precautionary principle, adopted by several international treaties such as the Cartagena Protocol on Biodiversity[19] and the Convention on Biological Diversity[20] as well as the European Union, which argues that decisions should be made and actions taken that err on the side of protecting health and the environment in situations characterized by scientific uncertainty.

Examples of differing world views are most easily cited from experiences with agricultural biotechnology. Although commercially deployed GE crops have generally had favorable economic outcomes for adopters of these crops, outcomes for farmers are heterogeneous because the social and economic effects depend not only on the fit of the crop variety to the environment, but also on the institutional support available to the farmer, such as access to credit, affordable inputs, extension services, and markets (NASEM, 2016b). With regard to access to affordable inputs and to markets, some people argue that the industrialization of agriculture through biotechnology may reduce the number of agents with economic access to agriculture (for example, inability of small farmers to compete with transnational enterprises) as well as decrease genetic diversity that can be achieved through plant-breeding programs and seed sharing at the grower level (Shiva et al., 2011; Vidal, 2011). In the context of biodiversity, some GE herbicide-resistant crops have been found to adversely affect populations of birds that feed on weed seeds due to such high levels of weed control (Gibbons et al., 2006).

Through an in-depth societal impacts analysis, social scientists have found that economic impacts of GE crops for different groups of farmers are mixed; that the political and regulatory context has significant impact on the ability of different groups to benefit; and that current private-sector control of GE crops, which is reinforced by the intellectual property system, reduces the ben-

---

[19]Cartagena Protocol on Biosafety. Available at https://bch.cbd.int/protocol/text. Accessed January 30, 2017.
[20]Convention on Biological Diversity. Available at https://www.cbd.int/convention/text. Accessed January 30, 2017.

efits of GE crops for poor farmers due to high seed costs and distributional constraints (Fischer et al., 2015). On a related note, some are concerned that industrial deployment of certain seed varieties over others may reduce the biodiversity of the food supply (that is, reduction in the seed varieties to be planted and cultivated worldwide) (Jacobsen et al., 2013). Concentration of the global transgenic seed market has been rapidly increasing, and food and agriculture are increasingly controlled by just a few companies which focus on profitable GE crop products (Bonny, 2014).

Another concern is the deployment of resources. Some argue that funding should be devoted to policy rather than technical solutions. For example, using data and historical analysis of GE crops and their effects, some scientists have argued that research funding currently available for the development of GE crops would be better spent in other areas (such as funding for nutrition, policy research, governance, and solutions originating closer to the local level) in order to sustainably provide sufficient food for the world's growing population (for example, IAASTD, 2009; Jacobsen et al., 2013). Scientists have also found that, so far, genetic engineering has not increased the yield potential of crops, though the technology has been used to reduce yield losses due to pests and research to improve nutrient use and increase the efficiency of photosynthesis is ongoing (NASEM, 2016b).

How important concerns about biotechnology are in comparison to the benefits provided depends not only on the interpretation of evidence, but also on an individual's and social group's perception of risk and technologies. Social science offers tools for understanding societal values and provides context for how disruptive technologies are viewed by different subgroups. Risk-perception theory points to different factors and cultural predispositions as to why people perceive risks differently including trust, risk and benefit distributions, controllability, familiarity, and world views (Slovic, 1987; Kahan et al., 2007; Kahan, 2012). Across multiple technological domains these factors affect risk perceptions in people, experts, and citizens alike.

Cultural groups that have been historically marginalized and do not hold as much power in society, such as women and underrepresented minorities, tend to rate risks higher and take more precautious attitudes toward technologies and risk than white men in the United States, even when education, income, and age are accounted for (Finucane et al., 2000; Kahan et al., 2007). An in-depth look at gender differences in response to environmental concerns also found that gender differences in risk perception seem to account for gender differences in worry about health-related environmental problems (McCright and Xiao, 2014). Cultural-cognition theory has been criticized, however, for its limitation to studies in Western, industrialized cultures, for its failings to account for more moderate positions, and for its tendency to blame the individual for their perceptions rather than to focus on risk reduction (Abel, 1985; Marris et al., 1998; van der Linden, 2016).

Other factors that contribute to perceptions of risks and benefits for technologies and their products have been studied and interpreted to form different theories and frameworks. For example, the psychometric paradigm focuses on identifying aspects of the technologies and the risks associated with them, such as whether or not these technologies and risks are dreaded, catastrophic, uncertain, voluntary, and novel, and how these factors affect risk perception and attitudes toward technologies (Fischhoff et al., 1978; Slovic, 1987). This theory suggests that risk perception would become more negative due to anxiety-provoking factors associated with biotechnology products such as uncertainty, involuntary exposure, unfamiliarity, uncontrollability, and catastrophic risk; this has been shown to be true in some studies that include genetic engineering in comparison to other technologies (Slovic, 1987; Marris et al., 1998). Controllability and familiarity have also decreased expert ratings of risks of potential future synthetic-biology products (Cummings and Kuzma, 2017).

There are also several sociological and cultural frameworks that emphasize the role of social factors in consumer attitudes toward products. Trust and confidence in social networks (for example, social groups, communities, extended families, and friends) and societal systems (that is, the market, the political system, the regulatory system, and news media) play an important role in perceptions of risk for products, especially when those risks are new, uncertain, or ambiguous (Rohrmann and Renn, 2000). They also influence people's reactions or behaviors in response to risk; for example, lack of trust in industry's ability to handle risk is associated with greater levels of political activism (Rohrmann and Renn, 2000). Returning to the example of agricultural biotechnology, a national public-perception study found that "trust of government to manage technology" was an important factor for influencing views about the balance of risks versus benefits of GE foods, which in turn affected decisions about acceptance and purchasing (Yue et al., 2015).

The committee notes that most of these theories and factors associated with risk perception are not unique to products of biotechnology. Similar factors can be observed in the perception of climate change by the general public in the United States, for example (Hansen et al., 2003; Kahan et al., 2012). The "deficit model," often promoted by natural scientists, presumes that there is a knowledge deficit in the public that can be corrected by giving more information and that, if members of the public are given the facts, they will support new technologies (Hansen et al., 2003). To the contrary, the field of public understanding of science has shown that, even with increased knowledge and information, a complex set of societal, political, individual, and cultural factors comes into play in people's perception of technologies and risk (Hansen et al., 2003; Kahan et al., 2012). According to the fields of public perception and risk communication, education is not likely to change public attitudes; scholars in these fields instead promote the idea of public deliberation, engagement, and communication to help increase technological understanding and mitigate unwarranted perceptions of risk deriving from social amplification and information asymmetries (for example, Thompson, 2011).

Impact on Regulation

Although societal benefits and societal values are not necessarily part of the process of assessing the technological risk associated with biotechnology products, they play an important role in the governance and oversight of biotechnology products, and the laws that society passes reflect societal values. As outlined in the next section, future biotechnology products have the capability to be much more complex than current products, and it is likely that these new products will have the promise of enhanced social benefits at the same time as being more controversial in terms of their use. The role of nontraditional developers may also play a strong role because many developers may not be as aware of the biotechnology regulatory system as current industrial players. The concerns around genome editing that have surrounded the advances in CRISPR-Cas9 are a preview of the scope and complexity of societal discourse that may surround future products of biotechnology (Baltimore et al., 2015; Ledford, 2016; NASEM, 2016b).

Finally, it is important to note that regulation is not the only means of governance and oversight. Codes of conduct, such as those developed in the DIYbio community (Kuiken, 2016), can also play an important role. Community and industry agreement on appropriate oversight frameworks and standards, even in the absence of explicit regulation, will be important for addressing how new products of biotechnology are evaluated.

FUTURE BIOTECHNOLOGY PRODUCTS

As a key element of the committee's charge, this section describes the future products of biotechnology and how their scope, scale, complexity, and tempo are accelerating. Examples of products that illustrate that change are given. As was stated in Chapter 1, for the purposes of the committee, biotechnology products are defined as products developed through genetic engineering or genome engineering (including products where the engineered DNA molecule is itself the "product," as in an engineered molecule used as a DNA information-storage medium) or the targeted or in vitro manipulation of genetic information of organisms, including plants, animals, and microbes. The committee also included products produced by such plants, animals, microbes, and cell-free systems or products derived from all of the above.

Increasing Scope, Scale, Complexity, and Tempo of Products

Many of the invited presentations the committee heard focused on the potential for increased scope, scale, complexity, and tempo of future products of biotechnology. These were also recurring themes throughout the committee's deliberations.

"Increased scope" means new types of biotechnology products that have not yet been handled by the U.S. regulatory system. Input from companies at the committee's information-gathering meetings and surveys conducted by the Woodrow Wilson Center for International Scholars revealed that future products of biotechnology are quite diverse and make use of a wide variety of host organisms—bacteria, fungi, plants, animals, and humans—to serve a large number of markets such as health, energy, environment, food, and personal care (Munnelly, 2016; Peck, 2016; Reed, 2016; Sewalt, 2016; Stanton, 2016).[21] Work in advanced academic laboratories and an industry report indicate growing interest in in vitro technologies (BIO, 2016). Plants that glow, yogurts that harbor biosensors, pigs that develop twice as much muscle, and microbial communities that may protect honey bees from parasitic mites are just a few possible future products of biotechnology in development.

"Increased scale" refers to the vast number of products (and variants thereof) that will enter the system as a consequence of advances in knowledge and technology. As the technologies for modifying genomes have expanded, so have the number of variants of prospective products. The increase in speed of the DBTL cycle (Gibson et al., 2010; Gill et al., 2016; Hutchison et al., 2016) has lowered the cost of creating variants of genes and pathways such that more prospective products can be screened for product viability.

The U.S. Environmental Protection Agency (EPA) has already noted an increase in the number of biotechnology products that are being submitted to it for regulatory purposes under the Toxic Substances Control Act (TSCA) (Figure 2-5). The number of microbial commercial activity notice submissions to EPA doubled from 2012 to 2013 and grew sharply in subsequent years. In addition, EPA identified that the newer submissions from algal-strain developers were from "companies that have had little or no experience with new substance review under TSCA" (EPA, 2015), which is in line with the observation that more actors with limited experience in the regulatory sphere are working on biotechnology-product development (see Box 2-2).

The number of products being submitted to the U.S. Department of Agriculture's Animal and Plant Inspection Service (USDA–APHIS) for field release has dropped from its peak in the late 1990s and early 2000s; however, the number of *gene constructs*[22] tested in those field releases is increasing. There has also been an increase since 2009 in the number of acres on which those releases are conducted (Table 2-2).

"Increased complexity" refers to a movement away from single-gene or single-pathway engineering using recombinant-DNA (rDNA) technology to the use of genome engineering to create multiplexed pathways and, at the extreme, engineered microbial communities for release in environments ranging from animal guts to large ecosystems. In considering biotechnology products, two biological "systems" are relevant: the host—the organism into which new material is introduced—and the source organism of the genetic material being introduced. Generally speaking, the majority

---

[21]See also Synthetic Biology Project. Available at http://www.synbioproject.org. Accessed October 11, 2016.

[22]A *gene construct* is the name used for a functional unit of DNA necessary for the transfer or the expression of a gene of interest. It includes the gene or genes of interest, a marker gene (to facilitate detection inside the organism), and appropriate control sequences as a single package.

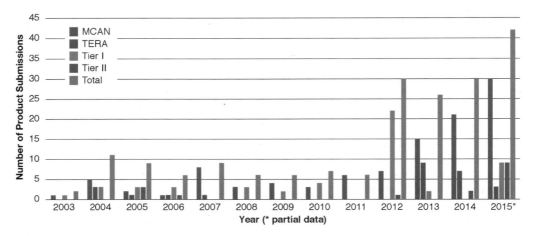

FIGURE 2-5 Increase in Toxic Substances Control Act (TSCA) biotechnology product submissions to the U.S. Environmental Protection Agency, 2003–2015.
NOTES: MCAN = Microbial commercial activity notice; TERA = TSCA experimental release application. A Tier I exemption requires certain certifications and recordkeeping. A Tier II exemption requires certain certifications and a notification to EPA and EPA review of specific physical containment and control technologies.
SOURCE: EPA (2015).

of biotechnology products in commerce as of 2016—such as crops genetically engineered to resist herbicides or insects—were the result of the transformation of a well-characterized host organism, such as corn or soybean, with a few genes from another source organism that code for a desired trait, such as herbicide resistance along with a selectable marker gene to permit selection for transformed plants (Figure 2-6, column A). Such organisms are easily compared against their nontransformed (that is, nonbiotechnology) counterparts in risk assessments. As biotechnologies have matured over time, new types of products are being developed that allow for the transformation of less well-characterized hosts. For example, new genome-editing technologies allow developers to make changes in genomes of nearly any host organism for which there is a genome sequence available, from microbes to insects to mammals (Figure 2-6, columns B and C; Reardon, 2016). Advances in biotechnology also allow the introduction of novel, synthetic gene sequences and the creation of consortia—a collection of genes derived from multiple unrelated sources—which can be inserted into a host (Figure 2-6, column D). The products that fall in the second, third, and fourth columns are not as easily compared against a nontransformed (nonbiotechnology) counterpart. The challenge this presents to the regulatory system is discussed in Chapter 4.

"Tempo" refers to the groups of similar new products that will predictably follow similar paths through the regulatory system following regulatory decisions about first-of-their-kind products. For example, in 2010 and 2011, USDA–APHIS reviewed a GE herbicide-resistant grass species transformed with biolistics instead of *Agrobacterium*; the agency decided that because the transformation did not occur through the use of a plant pest and the grass species was not classified as a plant pest, the grass did not qualify as a regulated article.[23] Because the grass was not intended for food and did not contain a pesticide, it did not fall under the regulatory purview of the U.S. Food and

---

[23]See letter from Michael C. Gregoire, Deputy Administrator, U.S. Department of Agriculture–Animal and Plant Health Inspection Service to Richard Shank, Senior Vice President, Scotts Miracle-Gro Company concerning confirmation of regulatory status of Kentucky bluegrass (July 1, 2011). Available at https://www.aphis.usda.gov/brs/aphisdocs/scotts_kbg_resp.pdf. Accessed January 9, 2017.

TABLE 2-2 Number of Releases, Gene Constructs, and Acres Authorized by the U.S. Department of Agriculture's Animal and Plant Health Inspection Service (USDA–APHIS) for Evaluation, 1987–2012

	Releases	Authorized Gene Constructs	Acres[a]
1987	11	5	—
1988	16	16	—
1989	30	30	—
1990	51	50	—
1991	90	89	—
1992	160	160	—
1993	301	306	948
1994	579	585	8,117
1995	711	710	62,394
1996	612	604	7,084
1997	763	761	23,817
1998	1,071	1,075	89,620
1999	983	1,005	56,959
2000	925	904	40,199
2001	1,083	1,083	54,195
2002	1,194	1,191	139,023
2003	813	810	24,713
2004	893	891	58,809
2005	955	956	99,510
2006	865	2,149	84,061
2007	932	4,920	45,931
2008	871	8,581	182,964
2009	751	16,650	166,315
2010	660	30,770	139,517
2011	792	35,186	235,226
2012	665	38,795	374,338
2013	602	50,963	368,384
2014	557	39,382	365,089
2015	467	46,214	447,631

[a]Records of the authorized planting acreages prior to 1993 incomplete.
SOURCE: USDA–APHIS (2017).

Drug Administration (FDA) or EPA either.[24] Between that decision in 2011 and December 2016, more than 40 GE plant products had been submitted to USDA–APHIS to determine if the product would fall outside the definition of a regulated article. Most have been determined to be outside the scope of USDA–APHIS's plant-pest authorities; however, the committee does not know which products have entered consumer markets.[25]

---

[24]See Chapter 3 for more discussion of the roles and responsibilities of the regulatory agencies and Appendix D for the Federal Food, Drug, and Cosmetic Act definition of *food*.

[25]Regulated Article Letters of Inquiry. Available at https://www.aphis.usda.gov/aphis/ourfocus/biotechnology/am-i-regulated/Regulated+Article+Letters+of+Inquiry. Accessed January 9, 2017.

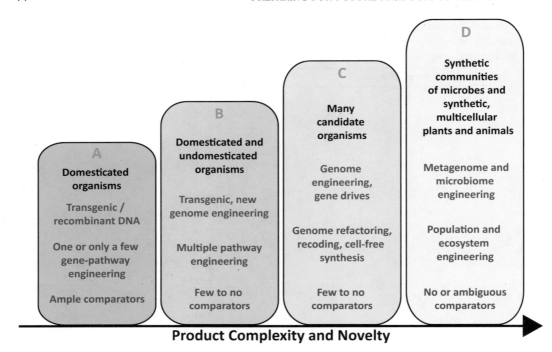

Product Complexity and Novelty

FIGURE 2-6 Characteristics of future biotechnology products, organized by similar levels of complexity in terms of types and number of organisms, genes and traits, and comparators involved.
NOTE: Products of biotechnology can be conceptualized as fitting into the depicted columns with the indicated characteristics, moving toward column D as a product increases in complexity and likelihood of providing new challenges for risk assessment.

The committee anticipated that a similar trend will be seen once the first genome-editing animal enters commerce. Advances in CRISPR-mediated genome editing have given rise to predictions of a wide range of precisely engineered animals including monkeys, mosquitoes, pigs, bees, cows, carp, dogs, ferrets, shrews, and chickens for an array of purposes including disease models, drug production, disease control, pets, food production, vector control, and behavioral studies (Reardon, 2016). Although it may be too soon to determine how many of these precisely engineered animals will enter commerce, developers who are planning to create such products of biotechnology are likely to be closely observing the regulatory path followed by the first successfully marketed animal.

Biotechnology Product Classes

The committee scanned the horizon for products emerging in the biotechnology space in the next 5–10 years. Horizon scanning is "a technique for detecting early signs of potentially important developments through a systematic examination of potential threats and opportunities, with emphasis on new technology and its effects on the issue at hand."[26] It is done worldwide to identify and understand the effects of new technologies; a well-established example is in the area of health technologies (Douw et al., 2003). The committee conducted its horizon-scanning exercise by inviting

---

[26]*Horizon scanning* in Overview of Methodologies. Organisation for Economic Co-operation and Development. Available at http://www.oecd.org/site/schoolingfortomorrowknowledgebase/futuresthinking/overviewofmethodologies.htm. Accessed December 18, 2016.

product developers to speak at the various data-gathering sessions; reviewing submitted public comments; reading scientific literature, popular press reports, and patents; consulting previous reports by the National Academies; searching publicly available iGEM projects; and checking information available on agency websites and crowdfunding websites. It also made use of the Synthetic Biology Database[27] curated by the Woodrow Wilson Center, which is focused on a subset of biotechnology products derived through the use of synthetic biology. After careful review of the products, the committee classified the products in order to better manage the task of describing them. All products within the scope of this report were grouped into three major classes: open-release products, contained products, and platforms. The following sections describe the qualities of products in each class and provide examples for products that regulators should expect to confront in the future.

Open-Release Products

This class includes all plants, animals, and microbes that have been engineered (either via rDNA techniques or genome engineering) that will be deliberately released in an open environment (Figure 2-7, Table 2-3). Anticipated future products include logical extensions of these products but also shift to products consisting of organisms whose genome could be largely synthetic, including both organisms with advanced genetic delimitation and those engineered to be capable of sustaining themselves in the environment. Additionally, it includes organisms that have gone extinct (or are close to being extinct) and may be revived (that is, deextinction).

The ability to sustain existence in the environment is a key change between existing products of biotechnology and some of the future ones anticipated in this class. As of 2016, most biotechnology products designed for open release into the environment were introduced into managed systems. For example, GE crops are grown in agricultural fields that are regularly tended and periodically harvested. Only a few deregulated GE crops (such as glyphosate-resistant alfalfa and virus-resistant papaya) are cultivated over more than one growing season, and these exceptions decline in productivity after a few years and are not designed to persist in the environment. However, some biotechnology products in development are being engineered to survive and persist in open environments with minimal or no management.

Furthermore, the types of environments in which a product may persist are likely to become more diverse. Plants and insects may be designed to continue in low-management systems such as forests, pastures, and cityscapes; microbes may be developed to persist in those environments as well as in mines, waterways, and animal guts. The committee anticipates that open-release products created to survive in a wide variety of environments will become more common in the next 5–10 years.

In Figure 2-7 and Table 2-3, biotechnology products are organized around their time to market (horizontal axis) and the family of the host organism, or lack thereof in the case of synthetic products (vertical axis). On the basis of its information-gathering efforts, the committee found that plant hosts will continue to be a dominant area for biotechnology-product development. At the time the committee was writing its report, genetically engineered traits were being introduced into crops other than just corn, soybean, and cotton (which were the most commonly engineered host plants in the 1996–2016 period), and traits besides insect resistance (conferred through the insertion of genes from *Bacillus thuringiensis*) and herbicide resistance were being engineered (NASEM, 2016b). Additionally, more techniques were being used along with or in place of rDNA technology. Genome editing was being used to knock out genes to create new traits, such as reducing browning of flesh in fruits and vegetables when exposed to oxygen, as was demonstrated in mushrooms (Waltz, 2016). Genome editing was also being used to introduce genes that improve disease resistance, for example

---

[27]Synthetic Biology Products and Applications Inventory. Available at http://www.synbioproject.org/cpi. Accessed October 11, 2016.

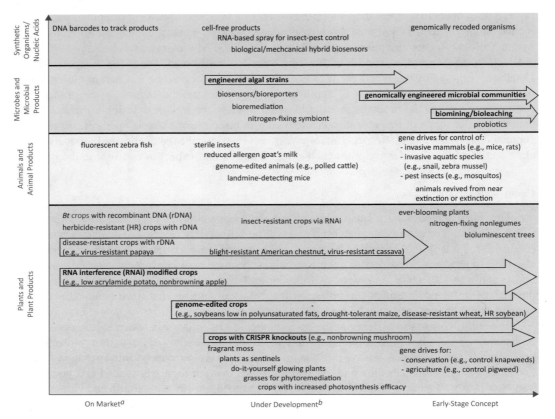

FIGURE 2-7 Market status of open-release products during the committee's study process.
NOTES: This figure diagrams market status of various open-release biotechnology products. Entries in bold are examples of products that the committee has identified as areas with high growth potential. Arrows indicate that the committee anticipated that similar products or products using the same transformation technology (for example genome editing or gene drives) are likely to be developed.
[a]"On Market" is equivalent to "In Use"; thus, products that have received regulatory approval but are not in use were not considered by the committee to be "On Market."
[b]"Under development" spans products from the prototype stage to field trials.

in wheat (Wang et al., 2014). Drought tolerance in corn (Shi et al., 2017) and more healthful oil quality in soybean (Haun et al., 2014) were being demonstrated through genome editing as well.[28] RNAi technology had already been used to introduce traits, including the reduction of browning in the flesh of apples and potatoes (NASEM, 2016b), and such products had cleared U.S. regulatory requirements. Scientists were using RNAi to create virus resistance in cassava, a staple crop in many African countries (Taylor et al., 2012). RNAi was also being used as a way different from rDNA-

---

[28]At the time the committee was writing its report, a canola variety with herbicide resistance had been commercialized by the company Cibus, which described the variety as developed using genome editing (Gocal, 2015). The resistance arose from a single nucleotide mutation in the BnAHAS1C gene selected for during an oligonucleotide-mediated genome-editing approach. However, Canadian regulatory documents note that, although the variety was developed using a genome-editing approach, Cibus "hypothesized that the single nucleotide mutation was the result of spontaneous somaclonal variation" rather than directly from the oligonucleotide-mediated editing. See Novel Food Information–Cibus Canola Event 5715 (Imidazolinone and Sulfonylurea Herbicide Tolerant). Available at http://www.hc-sc.gc.ca/fn-an/gmf-agm/appro/canola-5715-eng.php. Accessed March 31, 2017.

TABLE 2-3 Market Status of Products Designed for Open Release in the Environment[a]

	Product Description	On Market[b]	Under Development[c]	Early-Stage Concept
Plants and Plant Products	*Bt* crops with recombinant DNA (rDNA)	✓		
	Herbicide-resistant crops with rDNA	✓	✓	
	Disease-resistant crops with rDNA	✓	✓	
	RNAi modified crops	✓	✓✓✓	✓✓✓
	Fragrant moss		✓	
	Do-it-yourself glowing plants		✓	
	Genome-edited crops		✓✓✓	✓✓✓
	Crops with CRISPR knockouts		✓✓✓	✓✓✓
	Grasses for phytoremediation		✓	
	Plants as sentinels		✓	
	Crops with increased photosynthesis efficiency		✓	
	Ever-blooming plants			✓
	Nitrogen-fixing nonleguminous plants			✓
	Bioluminescent trees			✓
	Plants with gene drives for conservation purposes			✓
	Plants with gene drives for agricultural purposes			✓
Animals and Animal Products	Fluorescent zebra fish	✓		
	Sterile insects		✓	
	Genome-edited animals (e.g., polled cattle)		✓	✓
	Reduced-allergen goat's milk		✓	
	Landmine-detecting mice		✓	
	Animals revived from near extinction or extinction			✓
	Animals with gene drives for control of invasive mammals			✓
	Animals with gene drives for control of insect pests			✓
Microbes and Microbial Products	Biosensors/bioreporters		✓	
	Bioremediation		✓	
	Engineered algal strains		✓✓✓	
	Nitrogen-fixing symbionts		✓	
	Probiotics			✓
	Genomically engineered microbial communities			✓✓✓
	Biomining/bioleaching			✓✓✓
Synthetic Organisms/Nucleic Acids	Cell-free products		✓	
	DNA barcodes to track products	✓	✓	
	RNA-based spray for insect-pest control		✓	
	Genomically recoded organisms			✓
	Biological/mechanical hybrid biosensors		✓	✓

✓✓✓ = an area the committee has identified as having high growth potential.

[a]The table reflects the market status of products at the time the committee was writing the report.

[b]"On Market" is equivalent to "in use"; thus, products that have received regulatory approval but are not in use were not considered by the committee to be "On Market."

[c]"Under development" spans products from the prototype stage to field trials.

mediated toxin expression to introduce insect resistance into corn; this product was deregulated by USDA–APHIS in 2015.[29]

---

[29]See Petitions for Determination of Nonregulated Status. U.S. Department of Agriculture–Animal and Plant Health Inspection Service. Available at https://www.aphis.usda.gov/aphis/ourfocus/biotechnology/permits-notifications-petitions/petitions/petition-status. Accessed December 18, 2016.

Further out on the horizon, research was under way to reengineer processes in crops, such as photosynthesis (NASEM, 2016b). In 2016, tobacco was engineered to serve as a proof of concept for photosynthesis improvement. The engineered change increased the speed at which tobacco recovered from overexposure to sunlight (Kromdijk et al., 2016), which in turn increased leaf carbon dioxide uptake and plant dry matter productivity by 15 percent.

The committee anticipated that, in addition to crops, plants engineered for nonagricultural purposes would become more common and that many of these would be designed to persist in the environment under low and no management conditions. For example, a GE American chestnut contained an introduced enzyme, oxalate oxidase, which was extracted from wheat to confer resistance to a blight that has killed about 4 billion chestnut trees in North America since the early 1900s (Zhang et al., 2013). Many years of research had already been undertaken on this tree, and the committee thought it likely that the product would be submitted for regulatory approval within the next 5 years. Given the pressures that U.S. forests face from insect and disease infestation, invasive species, and the effects of climate change (Potter and Conkling, 2016), the committee thought it likely that more tree species will be engineered to resist such stressors.

Plants may also be engineered for biosecurity purposes. Scientists have engineered switchgrass and creeping bentgrass to degrade toxic munitions compounds from the soil in live fire–training ranges to prevent the toxins from leaching into groundwater (Zhang et al., 2017). Plants were also being engineered to serve as sentinels of environmental contamination (Kovalchuk et al., 1998; Kovalchuk and Kovalchuk, 2008).

Plants engineered for biosecurity purposes and trees represent examples of plants that will often be released into environments in which there is little or no management. The committee concluded that agricultural crops will continue to comprise the bulk of biotechnology plants but that more plants designed for little or no continual management will be developed than had been the case before 2016.

One other general type of open-release plant that the committee thought would become more common is one that is not designed for an agricultural or environmental purpose. Instead, the point is to appeal to consumers. The glowing plant mentioned earlier in the chapter is one such example. The committee heard that the product developer of that plant was also working on developing plants that are always in bloom, caffeinated apples, fragrant moss, and mosquito-repelling ivy (Evans, 2016).

As of 2016, few animals had been engineered for open release into the environment, but the committee anticipated that more such products would be developed in the next 5–10 years. At the time the committee was writing its report, U.S. regulators were already seeing insects transformed with rDNA technology that were created for open release. Two insect species had been engineered thus far, using two different approaches for controlling insect populations. A mosquito species (*Aedes aegypti*) was engineered to prevent the survival of all offspring and released-engineered adults without specialized treatment in laboratory conditions (Oxitec, 2016). The use of this strategy for biocontrol requires the repeated release of engineered adult male mosquitoes that have been reared in a laboratory to serve as breeding stock for wild female mosquitoes. Because the male mosquitoes (and subsequent offspring) are not designed to persist in the environment, it is anticipated that this intervention will have a limited environmental footprint beyond reducing the population of the *Aedes aegypti* species. A similar concept was being applied to control the population of diamondback moths; however, in this approach, only engineered males (and any resulting male offspring) can survive to adulthood. Over time, the balance of males to females shifts to the point that the population of moths would decline (Harvey-Samuel et al., 2015). As of February 2017, engineered diamondback moth and *Aedes aegypti* mosquito had completed contained trials and were being readied for environmental release in field tests in the United States.[30] Regulators could expect to see in the future variations of these biocontrol concepts applied to other insect lines, and possibly to mammals (such as invasive rodents) (Campbell et al., 2015).

Changes will also be introduced into livestock, which live in an open environment, though typically under conditions with regular human management and intervention. One example is an introduced trait that makes horned animals hornless. The trait has been demonstrated in cattle. Via TALENs, a naturally occurring polled allele has been isolated from hornless variants that are common in beef breeds and added to embryos from dairy breeds; the research resulted in two hornless calves (Carlson et al., 2016). Another biotechnological change that will likely be made to livestock animals is one that reduces the presence of allergens. As an example, scientists in China have reportedly modified goats to produce allergen-free milk (Zhu et al., 2016) by knocking out the whey protein that is the most common allergen for humans and knocking in a whey protein more similar in composition to human milk.

As with plants, animals may be modified for biosecurity purposes. Small mammals with an acute sense of smell can sniff out landmines without detonating them; giant African pouched rats have already been trained for this purpose. Mice have been engineered to have an odorant receptor that is particularly sensitive to explosives (D'Hulst et al., 2012), and they may transition from laboratory experiments to active identifiers of landmines in the next 5–10 years. Also as with plants, it is possible that more animals will be engineered to appeal to consumers. Engineered fluorescent zebra fish have been on the market since 2003. These fish were initially created for use in a research setting and then marketed to the general public (Nagare et al., 2009). The committee would expect other such novelty products to be developed, particularly in the pet market.

Biotechnology may be used to reintroduce extinct animals, or at least animals that are genetically similar to those that have gone extinct. Research is under way, for example, to use CRISPR genome editing to engineer elephant cells with mammoth versions of genes potentially involved in cold tolerance as a possible preamble to resurrecting the mammoth or creating an Asian elephant able to survive in cold temperatures (Callaway, 2015; Shapiro, 2015). Efforts are also ongoing for deextincting the passenger pigeon (Biello, 2014). The committee presumed that such animals, once approved by regulatory agencies, would be introduced into the environment under minimal or no management conditions.

Plants and animals with gene drives is a subclass of organisms that the committee also thought would be an area of growth in the biotechnology-product space in the next 5–10 years. A *gene drive* is a system of biased inheritance in which the ability of a genetic element to pass from a parent to its offspring through sexual reproduction is enhanced. Thus, the result of a gene drive is the preferential increase of a specific genotype that determines a specific phenotype from one generation to the next with the intention to spread throughout a population (NASEM, 2016a). Gene-drive mechanisms that have been explored for use in plants are for the control of knapweed for conservation purposes and for the control of pigweed in agricultural fields (NASEM, 2016a). In terms of animal applications, gene-drive mechanisms are being developed to control populations of the mosquito species *Culex quinquefasciatus* (which is a vector for avian malaria) and populations of non-native mice on islands (which negatively affect the habitats and ecosystems necessary for native species to thrive) (NASEM, 2016a). A future potential application is the use of a gene drive to spread disease resistance through a population of snails to prevent the continued transmission of schistosomiasis (Tennessen et al., 2015). Application of this technology to other invasive species

---

[30]Sterilized pink bollworm with a genetically engineered fluorescent marker has been field tested in Arizona since 2006. The marker allows for easy identification that the insect is sterile. Sterility in the insect has been achieved by irradiation, not genetic engineering.

has also been discussed in the popular press (Langin, 2014), yet it is unclear how many of these suggestions are being further developed.

As with animals, few microbes engineered for open release into the environment had been developed and successfully approved by regulatory agencies as of 2016. However, efforts have been under way for many years to genetically engineer microbes destined for the environment for a number of applications, including bioremediation (Cases and de Lorenzo, 2005) and as environmental biosensors (Xu et al., 2013). Such products were envisioned as future products of biotechnology in the 1986 Coordinated Framework for Regulation of Biotechnology. In the 1980s and early 1990s, limited field trials occurred with live GE organisms for such purposes as frost prevention and pest control. Recombinant biopesticides briefly formed a niche market in the early 1990s prior to development of insect-resistant transgenic crops. The failure of product advancement to the marketplace for engineered microbes may be attributed in some cases to a lack of performance of the product against expectations rather than evidence of failed safety tests (Wozniak et al., 2012). Other views posit alternative explanations for the lack of advancement of engineered microorganisms to commerce, including sentiment against GE organisms bringing the field of bioremediation to a standstill (de Lorenzo et al., 2016) and the inability to patent "non-novel" bioreporter technologies creating a disincentive to private-sector investment (Xu et al., 2013). An engineered bacterium, *Pseudomonas fluorescens* HK44, was the first GE microorganism to be field released for subsurface soil bioremediation of polycyclic aromatic hydrocarbons such as naphthalene and salicylate using bioluminescence. Despite promising results under a range of conditions, few applications of *Pseudomonas fluorescens* HK44 in relevant ecosystems have been implemented, "primarily due to legislative restrictions encompassing the use of genetically engineered microorganisms and their environmental release" (Trögl et al., 2012).

However, a *Pseudomonas putida* strain genetically engineered for aerobic bioremediation of 1,2,3-trichloropropane, a recalcitrant chlorinated hydrocarbon used as an industrial solvent, paint remover, and cleaning agent among other uses, was recently created as an attractive option for groundwater decontamination (Samin et al., 2014). Although the engineered strain performed well in a bioreactor with 1,2,3-trichloropropane as the only organic carbon source, its prospects for use as continuous bioremediation of 1,2,3-trichloropropane in contaminated environments remain untested. Because single-strain bioremediation approaches may be vulnerable to slow growth or high decay rates caused by reactive side products, the use of microbial consortia that could stimulate growth rates by cross-feeding or remove reactive dead-end metabolites is envisioned as a possible strategy to mitigate these vulnerabilities (Samin et al., 2014; Jia et al., 2016; Lindemann et al., 2016). Such anticipated environmental experiments of large-scale, self-propagating bioremediation approaches aimed to reduce the impacts of human-made pollution may be supported by the notion that "assuming a reasonable risk is preferable to the sure disastrous effect of inaction," given the prospect of increasing environmental and ecosystem degradation (de Lorenzo et al., 2016).

Synthetic biology holds a great deal of potential for microbes in open environments, an area that the committee sees as gaining momentum. Despite the historical challenges discussed above, this area is very active in current research (OSTP, 2012) and includes research to characterize and manipulate the microbiomes of essentially any life form or environment of interest. Prototype engineered biosensors already exist to traverse mammalian guts and "record" events of interest to which the microbe was exposed (Kotula et al., 2014). Similar systems are following in areas such as pollinator health (Kwong and Moran, 2016). Many researchers (Fredrickson, 2015; Jia et al., 2016) and a number of iGEM teams (iGEM, 2012, 2013, 2015a,b) have worked to establish stable synthetic consortia of microorganisms—and the biological principles behind their establishment and maintenance—that could be used as the bases of a wide variety of future applications. At the time the committee was writing its report, product developers were working to create engineered consortia of microorganisms to market as potential new products for open release for a broad range of markets including mining and human and plant nutrition.

Biomining involves the use of microorganisms to extract rare and base metals from minerals and ore. For example, the bacterium *Acidithiobacillus ferrooxidans* and relatives are able to assist with bio-oxidation and bioleaching of many types of minerals for mining. Biotechnology is being applied to enhance these processes, and research has been conducted to engineer microbes for increased redox potential and leaching rates (Brune and Bayer, 2012). Researchers and companies, such as Universal BioMining, are working on synthetic-biology techniques to create inoculants containing extremophiles with targeted genetic alterations designed for metal extraction (DaCunha, 2016). When deployed at scale in the field, these inoculants could greatly improve the capture of valuable metals such as gold and copper from the increasing supply of low-grade ore, while simultaneously reducing the environmental effects caused by traditional mining practices.

An invited speaker (Cumbers, 2016) described to the committee a series of small company efforts where open-release biotechnology products containing synthetically engineered microorganisms for the human gut are envisioned for enriched foods, medical purposes, and lifespan elongation. Microbial products that are genetically engineered are also under development for plant microbiomes. The clearest example is the manipulation of nitrogen fixation in heterologous prokaryotes (Smanski et al., 2014). One product concept is to implement the cluster in plant-associated microbes or, conversely, in microbes that the plant will selectively internalize. (The concept could also be implemented as a transgenic manipulation of the plant or a specific compartment in the plant.) Other applications currently under development are engineered bacterial strains that secrete double-stranded RNA (dsRNA) that can be applied to crop plants topically. The dsRNA produced by the bacteria is designed to serve as a crop protection agent, causing harm to pest insects that consume the treated plant (Killiny et al., 2014). Future microbial-consortium products of biotechnology, whether for bioremediation, biomining, or nutrition, present substantial challenges to regulators given their complexity, lack of comparators to nonbiotechnology products, and lack of predictive risk-assessment pathways available to evaluate their impacts and safety (see Chapter 4).

Advances in DNA synthesis and assembly technologies have created the possibility of engineering organisms whose genome is substantially altered and may consist largely of DNA sequences that have been chemically synthesized (Boeke et al., 2016; Hutchison et al., 2016). For example, work from the research groups of Farren Isaacs, George Church, and others have produced a family of novel prokaryotes (Isaacs et al., 2011; Lajoie et al., 2013; Mandell et al., 2015; Napolitano et al., 2016; Ostrov et al., 2016). Through advanced genome-engineering tools, an organism's genome has been recoded to change its organization (Isaacs et al., 2011; Lajoie et al., 2013) or to alter fundamentally how it codes and decodes information. Some variations repurpose codon usage without adding a requirement for new amino acids (Isaacs et al., 2011; Lajoie et al., 2013; Rovner et al., 2015; Napolitano et al., 2016; Ostrov et al., 2016), and other strategies fundamentally change codon usage and add new amino acid requirements (Mandell et al., 2015; Rovner et al., 2015). Each of these options results in genomically recoded organisms (GROs) with increased genetic isolation from other prokaryotes in the environment. While GROs are still in the early stages of development for research purposes, the committee can imagine open-release applications of such organisms for agricultural, bioremediation, and nutritional (probiotic) purposes. Their genomic isolation would prevent meaningful gene flow to or from the organism (Lajoie et al., 2013; Ma and Isaacs, 2016) and some of these recoding operations can be further modified to create highly effective kill switches, which can be leveraged to allow for tightly controlled environmental releases. No open-release products are being currently developed (or at least not yet disclosed), but one could anticipate the movement of these into regulatory purview in the future. It is much more likely that regulators will first see GROs for use as a contained product.

Other examples of synthetic products currently under development are synthetic RNAi to be used as pesticides and DNA barcodes to track products through the manufacturer pipeline. RNAi sprays, produced via a cell-free expression system to protect crops from insect pests, are under development and being tested in small greenhouse trials, with one demonstration showing the protection of potato plants from the Colorado potato beetle for up to 28 days (San Miguel and Scott, 2016). DNA barcodes have just begun to be utilized by the U.S. military to track small mechanical parts, such as bolts, to counter their rising concern over counterfeit parts of low quality (Mizokami, 2016). In this application, a DNA sequence is applied to a mechanical part using an epoxy ink. The novelty of such a tracking device is that it can be applied to small components of a system without impeding their function.

Contained Products

A second major class of products is those that are largely contained, that is, used in industrial fermentation or produced in other sealed environments such as laboratories or ponds. Organisms of many genera are used in fermenters to produce commodity chemicals, fuels, specialty chemicals or intermediates, enzymes, polymers, food additives, and flavors. When considering the laboratory as a contained environment, then many examples of transgenic animals from vendors are widely used today for research and development. Because performing biotechnology in contained environments allows higher control over the choice of host organism, systems with advanced molecular toolboxes are already in high use. As above, possible future biotechnology products captured in Figure 2-8 and Table 2-4 are organized around their time to market (horizontal axis) and the family of the host organism, or lack thereof in the case of synthetic products (vertical axis).

On the basis of its information-gathering efforts, the committee concluded that future biotechnology products that are produced in contained environments are more likely to be microbial based or synthetically based rather than based on an animal or plant host. However, the committee did identify a few animal and plant products, and they or variants thereof may become more common in the next 5–10 years. The CRISPR-edited mushroom, described in the open-release section above, can also be cultivated as a contained product in a laboratory or greenhouse setting (Waltz, 2016). An animal example approved by the regulatory agencies when the committee was writing its report was the GE salmon, which contains a gene insertion that speeds the pace at which the fish grows to market size. GE salmon are considered to be "contained" as a condition of regulatory approval because they are restricted to growth in specific land-based facilities and are prohibited from being grown in ocean net pens.[31] Another example of an existing animal product is laboratory animals, many of which are designed to have genes knocked in or out for experimental purposes, such as mini-swine (F. Li et al., 2014) or dogs (Zou et al., 2015) engineered for research purposes.[32] It is not clear how many follow-on product concepts are planned in this space by developers.

Other products that are not on the market yet, but that the committee thought could be commercialized in the near future, are polymers produced by plants for industrial use—for example, silk and collagen (reviewed by Van Beilen and Poirier, 2008)—and animal products derived from animal cells rather than from animals themselves. The committee heard from product developers working to create hamburgers by editing and expanding cultures of muscle cells in the laboratory (Datar, 2016; Shigeta, 2016). Leather from animal proteins expressed in skin cells has also been cre-

---

[31]U.S. Food and Drug Administration. AquAdvantage Salmon Fact Sheet. Available at http://www.fda.gov/Animal Veterinary/DevelopmentApprovalProcess/GeneticEngineering/GeneticallyEngineeredAnimals/ucm473238.htm. Accessed December 19, 2016.

[32]It is possible that such animals could one day be released into open environments for use as pets. It was reported in 2015 that an institute in China was considering selling as pets pigs that had been genome-edited for use as models for human disease (Cyranoski, 2015).

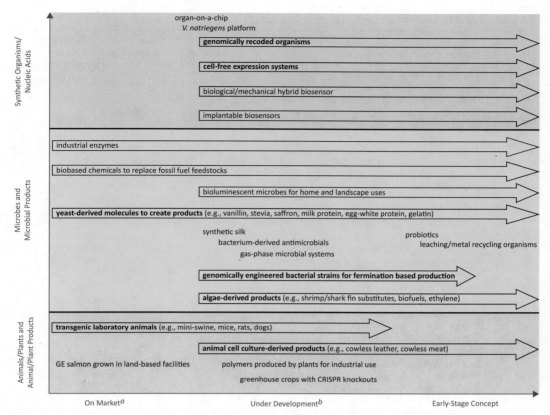

FIGURE 2-8 Market status of contained products during the committee's study process.
NOTES: This figure diagrams the type of technology and market status of various contained biotechnology products. Entries in bold are examples of products that the committee has identified as areas with high growth potential. Arrows indicate that the committee anticipated that similar products or products using the same transformation technology (for example genome editing or gene drives) are likely to be developed.

[a]"On Market" is equivalent to "In Use"; thus, products that have received regulatory approval but are not in use were not considered by the committee to be "On Market."

[b]"Under development" spans products from the prototype stage to field trials.

ated, and companies in this area are expanding and preparing to bring their products to the market in the near future (Shontell, 2016).

The committee also anticipated that replacements for products traditionally sourced from animals will be more and more likely to come from microbes such as yeasts or algae. In presentations by product developers, the committee heard about yeast strains developed to produce various products using traditional fermentation methods (Datar, 2016; Shigeta, 2016). A popular area of development is the creation of food products such as milk proteins to be combined with sugars and oil for the creation of vegan milk and cheese that are biologically the same as the animal sources of these products (Gertz, 2014; Bowler, 2016; Datar, 2016). A few other examples of food or flavor products being made in this way are egg whites (Datar, 2016; Shigeta, 2016), stevia,[33] gelatin (Duan et al.,

---

[33]In early 2016, Cargill, Inc., submitted a generally-recognized-as-safe exemption claim to FDA for steviol glycosides from *Saccharomyces cerevisiae* expressing steviol glycoside biosynthesis pathway (Cargill, 2016). In May 2016, FDA responded that it had no questions about Cargill's conclusion that the steviol glycosides were generally recognized as safe, based on information provided by Cargill and other information available to the agency (FDA, 2016).

TABLE 2-4 Market Status of Contained Products[a]

	Product Description	On Market[b]	Under Development[c]	Early-Stage Concept
Animals/Plants and Animal/Plant Products	Transgenic laboratory animals (mini-swine, mice, rats, dogs)	✓	✓✓✓	
	Genetically engineered salmon grown in land-based facilities	✓		
	Animal cell culture–derived products (e.g., cowless leather and cowless meat)		✓✓✓	✓✓✓
	Polymers produced by plants for industrial use		✓	
	Greenhouse crops with CRISPR knockouts		✓	
Microbes and Microbial Products	Industrial enzymes	✓	✓	✓
	Biobased chemicals to replace fossil fuel feedstocks	✓	✓	✓
	Bioluminescent microbes for home and landscape uses		✓	✓
	Yeast-derived molecules to create products (e.g., vanillin, stevia, saffron, egg whites, milk protein, gelatin)	✓	✓✓✓	✓✓✓
	Synthetic silk		✓	
	Bacterium-derived antimicrobials		✓	
	Genomically engineered bacterial strains for fermentation-based products		✓✓✓	
	Gas-phase microbial systems		✓	
	Algae-derived products (e.g., substitute for shark fins and shrimp, biofuels, ethylene)		✓✓✓	✓✓✓
	Probiotics			✓
	Leaching/metal recycling organisms			✓
Synthetic Organisms/Nucleic Acids	Organ-on-a-chip		✓	
	V. natriegens platform	✓	✓	
	Genomically recoded organisms		✓✓✓	✓✓✓
	Cell-free expression systems		✓✓✓	✓✓✓
	Biological–mechanical hybrid biosensor		✓	✓
	Implantable biosensors		✓	✓

✓✓✓ = an area the committee has identified as having high growth potential.

[a]The table reflects the market status of products at the time the committee was writing the report.

[b]"On Market" is equivalent to "in use"; thus, products that have received regulatory approval but are not in use were not considered by the committee to be "On Market."

[c]"Under development" spans products from the prototype stage to field trials.

2011), and vanillin (Hansen et al., 2009). The committee expects to see growth in the number and market acceptance of such food products as they are being marketed as more sustainable and cruelty-free products.[34] Other applications include the production of materials, such a silk (Tokareva et al., 2013), which some developers are working to optimize to create improved versions of this material (Seltenrich, 2015). Similarly, strains of algae have been engineered for use in fermentation processes to produce food products and other important industrial chemicals. Products such as vegan shrimp or shark fins made from collagen expressed by GE algae and combined with other ingredients are under development (Davis, 2015; Bryce, 2016; Shigeta, 2016). Other GE algae strains are being utilized

---

[34]See, for example, Memphis Meats, About Us: Better Meat, Better World, available at http://www.memphismeats.com/about-us, and Clara Foods: Egg Whites Without Hens at http://www.new-harvest.org/clara_foods. Accessed January 8, 2017.

to produce ethylene (Xiong et al., 2015), for use in making downstream products such as plastic, polyester, and PVC pipes, and for biofuel production (Radakovits et al., 2010).

The committee also expects to see microbial organisms with bioluminescent capabilities engineered to glow brighter for human use as novelty items in the home (Lafrance, 2015; Lombardo, 2015) or as greener light sources in urban settings (Marcellin, 2016). These products are designed for use in contained settings and not intended for consumers to release into the environment. Other microbial products are bacterial strains engineered to extract metal compounds and improve recovery from mining ores in a contained reactor (Schippers et al., 2014) or to produce a wide variety of chemical compounds such as antimicrobials or industrial enzymes. The regulators have long seen such products, with the classic example being insulin for medical purposes. A more recent, synthetic biology example is being deployed against traditional gas-phase fermentation organisms (already in use at pilot or demonstration scale for ethanol production from CO or CO_2) to expand the chemical targets beyond ethanol, acetate, and butanediol (Liew et al., 2016). Engineering bacterial strains to improve their production abilities is an area that the committee anticipates will grow. Many developers are creating production strains for their own production use or for selling to partners (see Table 2-5 for examples of such companies). There are numerous sources for production hosts, but one specific and recently deployed example to accelerate the DBTL cycle is *Vibrio natriegens*, a marine bacterium with a generation time of 10 minutes under optimal conditions (Lee et al., 2016; Weinstock et al., 2016). This generation time is much shorter than other production hosts and will allow developers to rapidly scale up new products and increase their yield. Future variations of this strain, or other bacteria species, should be expected by the regulators in the coming years.

On the far end of the spectrum of chassis modifications are the GROs mentioned in the previous section, which have engineered features that allow for tightly controlled release applications. However, it seems more likely that their initial exposure to the regulatory agencies will be in contained operations where the product value can justify the novel hosts or the production of novel classes of bioproducts (for example, novel sequence-defined synthetic polymers or materials comprising many instances of nonstandard amino acids), or where intellectual-property security or biocontainment in the event of unintended release is desired to be high and enhanced barriers for genetic isolation are preferred. Further research and development of GROs will provide developers the ability to better predict how adding in sequences for the production of novel products will be received by the chassis organism and likely contribute to an accelerated DBTL cycle (Way et al., 2014). It would seem that an anticipatory opportunity for agencies and primary technology developers would lie in establishing rubrics for demonstrating the basic biosafety profiles of novel production strains. Doing this might have two benefits, one in clarifying data packages or dossiers (that is, the studies that form the knowledge base to inform a risk assessment) for entities bringing the initial submission of a novel host to a regulatory agency, but also establishing data reuse protocols or other considerations which would incentivize others to follow in adoption of new host chassis.

There appears to be growing interest in cell-free expression systems, but it is not clear at this juncture what challenges would be present at an agency review.[35] Such systems can be utilized to produce research tools such as proteins for microarrays (Zarate and Galbraith, 2014) as well as RNA (J. Li et al., 2014). Other research entities are working on a number of biotechnology projects involving new materials, highly engineered biosensors (though these are likely to be used as medical devices and are outside the purview of this report), and engineered microbial consortia that will eventually leave the research setting, potentially making their way to the regulatory agencies.

An example of a biomolecular robotic sensing device funded by the U.S. government is Cyberplasm (Roberts et al., 2015). Cyberplasm integrates engineered bacteria, yeast, mammalian

---

[35]The update to the Coordinated Framework notes that "nucleic acids produced via cell-free synthesis are used for pesticidal purposes, these products are regulated by EPA/OPP" (EOP, 2017:Table 2).

TABLE 2-5 Examples of Companies Developing Engineered Microbial Strains

Company	Product
Aequor, Inc.	Engineered marine microbe for antifouling and antibiofilm
Caribou Bioscience	Precision cell engineering
Chain Biotechnology Ltd.	Microbial hosts (chassis) for engineering anaerobic bacteria
DNA 2.0	Gene synthesis—tools provider
Enevolv, Inc.	Engineered microbes: bacteria, yeast, algae
Gen9[a]	Gene synthesis
Ginkgo Bioworks	Engineered microorganisms
Greenlight Biosciences	Cell-free bioprocessing technology
Molecular Assemblies	DNA synthesis
Muse Biotechnologies, Inc.	Strain engineering
Oligos Biotech	Engineered fungus
Pareto Biotechnologies	Polyketide pathways
Syngulon	Bacteriocin engineering
Synpromics	Synthetic promotors for gene expression
Synthetic Genomics	Advanced genomic engineering: microbial cell lines, DHA Omega-3, Astaxanthin
Teselagen	Combinatorial gene design and editing
Twist Bioscience	Gene synthesis on silicon
Zymergen	Strain improvement

[a]In January 2017, Ginkgo Bioworks acquired Gen9.
SOURCE: BIO (2016).

cells, and cell parts to undertake device-like functions capable of sensing and treating pathogens or chemicals within plants and animals or for other functions involving environmental sensing and remediation (for example, Ayers et al., 2010). Cell parts for detection, signaling, motion, and delivery would be integrated into a biogel matrix to mimic the movement of the sea lamprey. Another example of a combination of mechanical and living cells is organ-on-a-chip systems for research purposes (Bhatia and Ingber, 2014).

Biotechnology "Platforms"

A final class of products of biotechnology includes those that are used as platforms in the creation of other biotechnology products. New tools are always in the development pipeline and often represent the fastest-to-market products of biotechnology. Tools include products that are traditionally characterized as "wet lab," such as DNA/RNA, enzymes, vectors, cloning kits, cells, library prep kits, and sequencing prep kits, and products that are "dry lab," such as vector drawing software, computer-aided design software, primer calculation software, and informatics tools. These two categories continue to meld as newer approaches are published or commercialized. For instance, cloning from native hosts can be replaced by automated systems taking genomic sequence and creating libraries of genetic elements by blending DNA synthesis, assembly protocols, and automation workflows. Developers can currently access these as complete packages offered by third parties or *à la carte* by making a customized flow in house. Some of these services are available as cloud-based laboratory resources.

One dimension of this is the scale on which it is taking place. For instance, Twist Bioscience and Ginkgo Bioworks disclosed a commitment to transact 100 million bases of DNA, reportedly representing about 10 percent of the global synthesis market (Segran, 2015). Companies specializing in new software tools to manage such workflows are now squarely in this part of the biotechnology value chain.

Another dimension of change is the customer base of such product offerings. Such kits, software, and even hardware no longer reside exclusively in academic or industrial settings, as discussed previously in this chapter. The growing collegiate competition community, iGEM, actively consumes and contributes to a growing body of "parts" and kits. Likewise, the DIYbio community is increasingly interested in tools for genome-scale engineering, and more economical hardware and automation is being developed in maker spaces. Community laboratories are being used more and more by the DIYbio community and for science education (for instance, Building with Biology from the National Science Foundation). Amino Labs is producing a benchtop biolab for home use, and similar products can be expected to follow suit in the coming years.

SUMMARY AND CONCLUSIONS

This chapter describes some of the technological, economic, and social trends that will likely drive the development of future products of biotechnology; outlines the changes in the scope, scale, complexity, and tempo of biotechnology products; and provides a detailed account of new products that are likely to emerge in the next 5–10 years. The committee reached the following broad conclusions regarding emerging trends and products of biotechnology.

Conclusion 2-1: The U.S. bioeconomy is growing rapidly; the scope, scale, complexity, and tempo of biotechnology products are increasing.

Factors contributing to this increase include advances in basic understanding of biological processes that enable biotechnology products to be designed for a wider range of applications, more efficient and robust technologies such as genome-editing techniques, the "standardization" of bioengineering components, and the decreasing costs of genome sequencing and gene synthesis (which in turn decrease the costs of DBTL cycles and enable more startup companies to enter this area) and increasing amounts and sources of funding such as crowdfunding. Indeed, these factors have fostered a thriving DIYbio community in the United States and student competitions to design biotechnology products. With so many new actors, tools, and resources involved, future products of biotechnology are poised to permeate all aspects of human endeavor.

Conclusion 2-2: Societal factors will continue to play an important role in the public debate regarding the safe and effective use of biotechnology products.

The committee notes that there are many competing interests, risks, and benefits regarding future biotechnology products. The United States and international regulatory systems will need to achieve a balance among these competing aspects when considering how to manage the development and use of new products of biotechnology. Many parts of society have concerns over the safety and ethics of various biotechnologies, while others see prospects for biotechnology to address challenging social and environmental issues. Biotechnology products that are on the horizon are likely to generate substantial public debate. For example, gene-drive technology, for which there have already been numerous studies and reports regarding its use, is a type of technological advance that will increase the amount of public debate and for which society will have to take a balanced approach among the interested and affected parties, developers, and scientists.

Conclusion 2-3: Many future biotechnology products will be similar to existing biotechnology products, but they may be created through new processes.

Biotechnology products that have become familiar—such as insect-resistant crops and products made with bacteria in fermentation processes—will continue to be developed, though the method of genetic transformation will likely change. New forms of genetic transformation that are faster and more specific than recombinant-DNA technology, such as CRISPR, will likely allow product developers to design, build, test, and learn from experiments and product development more quickly in the next 10 years than has been the case in the last two decades. Increases in the speed and efficiency of DBTL cycles, along with an increase in the number of actors in the biotechnology space, will translate into the availability of more biotechnology products that are similar to the biotechnology products of 2016 but are transformed through processes other than recombinant DNA.

Conclusion 2-4: Some future products of biotechnology may be wholly unlike products that existed in 2016.

The increased capabilities to transform genomes afforded by advances in genomic engineering allow product developers to expand the number and kinds of modifications in future biotechnology products. The committee anticipates growth in the genetic transformation of microbes (such as yeast, algae, and bacteria) in contained systems to produce products such as chemicals and biofuels. It also expects the development of communities of microbes that are created from synthetic DNA and microbial communities formed from the combination of DNA from a number of different microbes; such communities may be designed for release in open environments to enhance nitrogen fixation by plants or for bioremediation use at contaminated sites. A much broader array of host organisms targeted for genetic transformation is also likely.

Conclusion 2-5: Novel biotechnology platforms will contribute to an increase in biotechnology products.

Biotechnology platforms—such as computational tools to improve efficiency, novel biotechnology kits to provide new actors the tools to become developers, improved capability to synthesize DNA and RNA, and increasingly automated systems—will proliferate and add to the number of and speed at which novel biotechnology products are developed.

REFERENCES

Abel, R.L. 1985. Blaming victims. Law & Social Inquiry 10(2):401–417.

Amiram, M., A.D. Haimovich, C. Fan, Y.S. Wang, H.R. Aerni, I. Ntai, D.W. Moonan, N.J. Ma, A.J. Rovner, S.H. Hong, N.L. Kelleher, A.L. Goodman, M.C. Jewett, D. Soll, J. Rinehart, and F.J. Isaacs. 2015. Evolution of translation machinery in recoded bacteria enables multi-site incorporation of nonstandard amino acids. Nature Biotechnology 33:1272–1279.

Ayers, J., N. Rulkov, D. Knudsen, Y.-B. Kim, A. Volkovskii, and A. Selverston. 2010. Controlling underwater robots with electronic nervous systems. Applied Bionics and Biomechanics 7(1):57–67.

Baltimore, D., P. Berg, M. Botchan, D. Carroll, R.A. Charo, G. Church, J.E. Corn, G.Q. Daley, J.A. Doudna, M. Fenner, H.T. Greely, M. Jinek, G.S. Martin, E. Penhoet, J. Puck, S.H. Sternberg, J.S. Weissman, and K.R. Yamamoto. 2015. A prudent path forward for genomic engineering and germline gene modification. Science 348(6230):36–38.

Bhatia, S.N., and D.E. Ingber. 2014. Microfluidic organs-on-chips. Nature Biotechnology 32(8):760–772.

Biello, D. August 29, 2014. Ancient DNA could return passenger pigeons to the sky. Scientific American. Available at https://www.scientificamerican.com/article/ancient-dna-could-return-passenger-pigeons-to-the-sky. Accessed January 8, 2017.

BIO (Biotechnology Innovation Organization). 2016. Advancing the Biobased Economy: Renewable Chemical Biorefinery Commercialization, Progress, and Market Opportunities, 2016 and Beyond. Available at https://www.bio.org/sites/default/files/BIO_Advancing_the_Biobased_Economy_2016.pdf. Accessed October 11, 2016.

Biomass R&D Board. 2016. Federal Activities Report on the Bioeconomy. Available at http://www.biomassboard.gov/pdfs/farb_2_18_16.pdf. Accessed August 10, 2016.

Boeke, J.D., G. Church, A. Hessel, N.J. Kelley, A. Arkin, Y. Cai, R. Carlson, A. Chakravarti, V.W. Cornish, L. Holt, F.J. Isaacs, T. Kuiken, M. Lajoie, T. Lessor, J. Lunshof, M.T. Maurano, L.A. Mitchell, J. Rine, S. Rosser, N.E. Sanjana, P.A. Silver, D. Valle, H. Wang, J.C. Way, and L. Yang. 2016. The genome project–write. Science 353:126–127.

Bonny, S. 2014. Taking stock of the genetically modified seed sector worldwide: Market, stakeholders, and prices. Food Security 6(4):525–540.

Bowler, J. August 29, 2016. This new dairy alternative is made from real milk proteins—but with no cows required. Science Alert. Available at http://www.sciencealert.com/this-new-dairy-alternative-is-made-from-real-milk-proteins-but-with-no-cows-required. Accessed January 8, 2017.

Brune, K., and T. Bayer. 2012. Engineering microbial consortia to enhance biomining and bioremediation. Frontiers in Microbiology 3:203.

Bryce, E. August 26, 2016. Synthetic prawns: A bid to make 'seafood' that's sustainable and slavery-free. The Guardian. Available at https://www.theguardian.com/environment/world-on-a-plate/2016/aug/26/synthetic-prawns-a-bid-to-make-seafood-thats-sustainable-and-slavery-free. Accessed January 8, 2017.

Callaway, E. 2015. Mammoth genomes hold recipe for Arctic elephants. Nature 521(7550):18–19.

Campbell, K.J., J. Beek, C.T. Eason, A.S. Glen, J. Godwin, F. Gould, N.D. Holmes, G.R. Howald, F.M. Madden, J.B. Ponder, D.W. Threadgill, A.S. Wegmann, and G.S. Baxter. 2015. The next generation of rodent eradications: Innovative technologies and tools to improve species specificity and increase their feasibility on islands. Biological Conservation 185:47–58.

Cargill. 2016. GRAS Exemption Claim for Steviol Glycosides from *Saccharomyces cerevisiae* Expressing Steviol Glycoside Biosynthesis Pathway. February 1. Available at http://www.fda.gov/downloads/Food/IngredientsPackagingLabeling/GRAS/NoticeInventory/ucm505151.pdf. Accessed January 8, 2017.

Carlson, D.F., C.A. Lancto, B. Zang, E.-S. Kim, M. Walton, D. Oldeshulte, C. Seabury, T.S. Sonstegard, and S.C. Fahrenkrug. 2016. Production of hornless dairy cattle from genome-edited cell lines. Nature Biotechnology 34(5):479–481.

Carlson, R. 2016. Estimating the biotech sector's contribution to the US economy. Nature Biotechnology 34(3):247–255.

Cases, I., and V. de Lorenzo. 2005. Genetically modified organisms for the environment: Stories of success and failure and what we have learned from them. International Microbiology 8(3):213–222.

Chambers, S., R. Kitney, and P. Freemont. 2016. The Foundry: The DNA synthesis and construction foundry at Imperial College. Biochemical Society Transactions 44:687–688.

Cohen, S.N., A.C.Y. Chang, H. Boyer, and R.B. Helling. 1973. Construction of biologically functional bacterial plasmids in vitro. Proceedings of the National Academy of Sciences of the United States of America 70:3240–3244.

Cong, L., F.A. Ran, D. Cox, S. Lin, R. Barretto, N. Habib, P.D. Hsu, X. Wu, W. Jiang, L.A. Marraffini, and F. Zhang. 2013. Multiplex genome engineering using CRISPR/Cas systems. Science 339(6121):819–823.

Cumbers, J. 2016. Future Emerging Open Release Products. Presentation to the National Academies of Sciences, Engineering, and Medicine Committee on Future Biotechnology Products and Opportunities to Enhance Capabilities of the Biotechnology Regulatory System, June 1, Washington, DC.

Cummings, C., and J. Kuzma. 2017. Societal risk evaluation scheme (SRES): Scenario-based multi-criteria evaluation of synthetic biology applications. PLOS ONE 12:e0168564.

Cyranoski, D. September 29, 2015. Gene-edited 'micropigs' to be sold as pets at Chinese institute. Nature 526(7571):18.

DaCunha, C. 2016. Universal Bio Mining: Transforming Mining with Synthetic Biology. Presentation to the National Academies of Sciences, Engineering, and Medicine Committee on Future Biotechnology Products and Opportunities to Enhance Capabilities of the Biotechnology Regulatory System, June 1, Washington, DC.

Datar, I. 2016. Small Business Perspectives from the 501(c)(3) Research Institute Advancing Cellular Agriculture. Presentation to the National Academies of Sciences, Engineering, and Medicine Committee on Future Biotechnology Products and Opportunities to Enhance Capabilities of the Biotechnology Regulatory System, June 27, San Francisco, CA.

Davis, N. November 14, 2015. Biotech bid to take shark off the menu and cut the fin trade. The Guardian. Available at https://www.theguardian.com/environment/2015/nov/15/shark-fin-soup-lab-grown-conservation-finning. Accessed January 8, 2017.

de Lorenzo, V., P. Marliere, and R. Solé. 2016. Bioremediation at a global scale: From the test tube to planet Earth. Microbial Biotechnology 9(5):618–625.

D'Hulst, C., M. Saparauskaite, and P. Feinstein. 2012. Generating a biosensor for the detection of landmines. Program Poster No. 815.09/FFF73. 2012 Neuroscience Meeting Planner. New Orleans, LA: Society for Neuroscience. Available at http://www.abstractsonline.com/Plan/ViewAbstract.aspx?sKey=5d17d281-d287-4859-8092-77a84caf35c9&cKey=7f76a7f8-fee2-4d64-925d-b88be0b87d90&mKey=%7b70007181-01C9-4DE9-A0A2-EEBFA14CD9F1%7d. Accessed December 18, 2016.

Doudna, J.A., and E. Charpentier. 2014. Genome editing. The new frontier of genome engineering with CRISPR-Cas9. Science 346(6213):1258096.

Douw, K., H. Vondeling, D. Eskildsen, and S. Simpson. 2003. Use of the Internet in scanning the horizon for new and emerging health technologies: A survey of agencies involved in horizon scanning. Journal of Medical Internet Research 5(1):e6.

Duan, H., S. Umar, R. Xiong, and J. Chen. 2011. New strategy for expression of recombinant hydroxylated human-derived gelatin in *Pichia pastoris* KM71. Journal of Agricultural and Food Chemistry 59(13):7127–7134.

EC (European Commission). 2012. The European Bioeconomy in 2030: Delivering Sustainable Growth by Addressing the Grand Societal Challenges. Available at http://www.epsoweb.org/file/560. Accessed August 10, 2016.

Eisenstein, M. 2016. Living factories of the future. Nature 531(7594):401–403.

El-Chichakli, B., J. von Braun, C. Lang, D. Barben, and J. Philp. 2016. Policy: Five cornerstones of a global bioeconomy. Nature 535(7611):221–223.

EOP (Executive Office of the President). 2017. Modernizing the Regulatory System for Biotechnology Products: An Update to the Coordinated Framework for the Regulation of Biotechnology. Available at https://obamawhitehouse.archives. gov/sites/default/files/microsites/ostp/2017_coordinated_framework_update.pdf. Accessed January 30, 2017.

EPA (U.S. Environmental Protection Agency). 2015. Biotechnology Algae Project. August 5. Available at https://www.epa.gov/ sites/production/files/2015-09/documents/biotechnology_algae_project.pdf. Accessed September 14, 2016.

Evans, A. 2016. TAXA. Presentation to the National Academies of Sciences, Engineering, and Medicine Committee on Future Biotechnology Products and Opportunities to Enhance Capabilities of the Biotechnology Regulatory System, June 27, San Francisco, CA.

EY. 2016. Beyond Borders 2016: Biotech Financing. Bountiful Harvest Leaves Biotech Well Prepared for Financial Winter. Available at http://www.ey.com/Publication/vwLUAssets/ey-beyond-borders-2016-biotech-financing/$FILE/ ey-beyond-borders-2016-biotech-financing.pdf. Accessed September 27, 2016.

FDA (U.S. Food and Drug Administration). 2016. Agency Response Letter GRAS Notice No. GRN 000626. May 27. Available at http://www.fda.gov/Food/IngredientsPackagingLabeling/GRAS/NoticeInventory/ucm512620.htm. Accessed January 8, 2017.

Finucane, M.L., P. Slovic, C.K. Mertz, J. Flynn, and T.A. Satterfield. 2000. Gender, race, and perceived risk: The "white male" effect. Health, Risk & Society 2:159–172.

Fischer, K., E. Ekener-Petersen, L. Rydhmer, and K.E. Björnberg. 2015. Social impacts of GM crops in agriculture: A systematic literature review. Sustainability 7(7):8598–8620.

Fischhoff, B., P. Slovic, S. Lichtenstein, S. Read, and B. Combs. 1978. How safe is safe enough? A psychometric study of attitudes towards technological risks and benefits. Policy Sciences 9(2):127–152.

Formas (Swedish Research Council for Environment, Agricultural Sciences and Spatial Planning). 2012. Swedish Research and Innovation Strategy for a Bio-based Economy. Available at http://www.formas.se/PageFiles/5074/Strategy_ Biobased_Ekonomy_hela.pdf. Accessed August 10, 2016.

Fredrickson, J.K. 2015. Ecological communities by design. Science 348(6242):1425–1427.

Gaj, T., C.A. Gerbach, and C.F. Barbas. 2013. ZFN, TALEN, and CRISPR/Cas-based methods for genome engineering. Trends in Biotechnology 31:397–405.

GBC (German Bioeconomy Council). 2016. Bioeconomy Policy (Part I): Synopsis and Analysis of Strategies in the G7. Available at http://bioeconomia.agripa.org/download-doc/64046. Accessed August 10, 2016.

Gertz, E. August 6, 2014. Can biohackers succeed at making "real vegan cheese"? Popular Science. Available at http://www. popsci.com/article/science/can-biohackers-succeed-making-real-vegan-cheese. Accessed January 8, 2017.

Gibbons, D.W., D.A. Bohan, P. Rothery, R.C. Stuart, A.J. Haughton, R.J. Scott, J.D. Wilson, J.N. Perry, S.J. Clark, R.J.G. Dawson, and L.G. Firbank. 2006. Weed seed resources for birds in fields with contrasting conventional and genetically modified herbicide-tolerant crops. Proceedings of the Royal Society B 273(1596):1921–1928.

Gibson, D.G., J.I. Glass, C. Lartigue, V.N. Noskov, R.-Y. Chuang, M.A. Algire, G.A. Benders, M.G. Montague, L. Ma, M.M. Moodie, C. Merryman, S. Vashee, R. Krishnakumar, N. Assad-Garcia, C. Andrews-Pfannkoch, E.A. Denisova, L. Young, Z.-Q. Qi, T.H. Segall-Shapiro, C.H. Calvey, P.P. Parmar, C.A. Hutchison, H.O. Smith, and J.C. Venter. 2010. Creation of a bacterial cell controlled by a chemically synthesized genome. Science 329(5987):52–56.

Gill, R.T., A.L. Halweg-Edwards, A. Clauset, and S.F. Way. 2016. Synthesis aided design: The biological design-build-test engineering paradigm? Biotechnology and Bioengineering 113(1):7–10.

Gocal, G. 2015. Non-transgenic trait development in crop plants using oligo-directed mutagenesis: Cibus' Rapid Trait Development System. Pp. 97–106 in *NABC Report 26 New DNA-Editing Approaches: Methods, Applications, & Policy for Agriculture*, A. Eaglesham and R.W.F. Hardy, eds. Ithaca, NY: North American Agricultural Biotechnology Council.

Grushkin, D., T. Kuiken, and P. Millet. 2014. Seven Myths & Realities about Do-It-Yourself Biology. Washington, DC: Woodrow Wilson Center for International Scholars.

Hansen, E.H., B.L. Møller, G.R. Kock, C.M. Bünner, C. Kristensen, O.R. Jensen, F.T. Okkels, C.E. Olsen, M.S. Motawia, and J. Hansen. 2009. De novo biosynthesis of vanillin in fission yeast (*Schizosaccharomyces pombe*) and baker's yeast (*Saccharomyces cerevisiae*). Applied and Environmental Microbiology 75(9):2765–2774.

Hansen, J., L. Holm, L. Frewer, P. Robinson, and P. Sandøe. 2003. Beyond the knowledge deficit: Recent research into lay and expert attitudes to food risks. Appetite 41(2):111–121.

Harvey-Samuel, T., N.I. Morrison, A.S. Walker, T. Marubbi, J. Yao, H.L. Collins, and L. Alphey. 2015. Pest control and resistance management through release of insects carrying a male-selecting transgene. BMC Biology 13(1):49.

Haun, W., A. Coffman, B.M. Clasen, Z.L. Demorest, A. Lowy, E. Ray, A. Retterath, T. Stoddard, A. Juillerat, F. Cedrone, L. Mathis, D.F. Voytas, and F. Zhang. 2014. Improved soybean oil quality by targeted mutagenesis of the fatty acid desaturase 2 gene family. Plant Biotechnology Journal 12(7):934–940.

Hayden, E.C. 2015. Tech investors bet on synthetic biology. Nature 527(7576):19.

Hutchison, C.A., III, R.Y. Chuang, V.N. Noskov, N. Assad-Garcia, T.J. Deerinck, M.H. Ellisman, J. Gill, K. Kannan, B.J. Karas, L. Ma, J.F. Pelletier, Z.Q. Qi, R.A. Richter, E.A. Strychalski, L. Sun, Y. Suzuki, B. Tsvetanova, K.S. Wise, H.O. Smith, J.I. Glass, C. Merryman, D.G. Gibson, and J.C. Venter. 2016. Design and synthesis of a minimal bacterial genome. Science 351(6280):aad6253.

IAASTD (International Assessment of Agricultural Knowledge, Science and Technology for Development). 2009. Agriculture at a Crossroads: Global Report. Washington, DC: Island Press.

iGEM (The International Genetically Engineered Machine). 2012. Modeling Microbial Consortia: The Auxotroph Approach, University of British Columbia. Available at http://2012.igem.org/Team:British_Columbia/Consortia. Accessed October 11, 2016.

iGEM. 2013. Engineering a Synthetic Microbial Consortium, Technische Universität Braunschweig. Available at http://2013.igem.org/Team:Braunschweig. Accessed October 11, 2016.

iGEM. 2015a. Synenergene, Team Amsterdam. Available at https://www.synenergene.eu/content/igem-2015-team-amsterdam. Accessed October 11, 2016.

iGEM. 2015b. Massachusetts Institute of Technology. Human Practices: State of the Art in Consolidated Bioprocessing. Available at http://2015.igem.org/Team:MIT/CBP. Accessed October 11, 2016.

Isaacs, F.J., P.A. Carr, H.H. Wang, M.J. Lajoie, B. Sterling, L. Kraal, A.C. Tolonen, T.A. Gianoulis, D.B. Goodman, N.B. Reappas, C.J. Emig, D. Bang, S.J. Hwang, M.C. Jewett, J.M. Jacobson, and G.M. Church. 2011. Precise manipulation of chromosomes in vivo enables genome-wide codon replacement. Science 333(6040):348–353.

Jacobsen, S.E., M. Sørensen, S.M. Pedersen, and J. Weiner. 2013. Feeding the world: Genetically modified crops versus agricultural biodiversity. Agronomy for Sustainable Development 33(4):651–662.

Jia, X., C. Liu, H. Song, M. Ding, J. Du, Q. Ma, and Y. Yuan. 2016. Design, analysis and application of synthetic microbial consortia. Synthetic and Systems Biotechnology 1(2):109–117.

Jinek, M., A. East, A. Cheng, S. Lin, E. Ma, and J. Doudna. 2013. RNA-programmed genome editing in human cells. Elife 2:e00471.

Juhas, M., and J.W. Ajioka. 2017. High molecular weight DNA assembly in vivo for synthetic biology applications. Critical Reviews in Biotechnology 37(3):277–286.

Kahan, D.M. 2012. Cultural cognition as a conception of the cultural theory of risk. Pp. 725–759 in Handbook of Risk Theory, S. Roeser, R. Hillerbrand, P. Sandin, and M. Peterson, eds. Dordrecht, Netherlands: Springer.

Kahan, D.M., D. Braman, J. Gastil, P. Slovic, and C. Mertz. 2007. Culture and identity-protective cognition: Explaining the white-male effect in risk perception. Journal of Empirical Legal Studies 4:465–505.

Kahan, D.M., E. Peters, M. Wittlin, P. Slovic, L.L. Ouellette, D. Braman, and G. Mandel. 2012. The polarizing impact of science literacy and numeracy on perceived climate change risks. Nature Climate Change 2(10):732–735.

Killiny, N., S. Hajeri, S. Tiwari, S. Gowda, and L.L. Stelinski. 2014. Double-stranded RNA uptake through topical application, mediates silencing of five CYP4 genes and suppresses insecticide resistance in Diaphorina citri. PLOS ONE 9(10):e110536.

Klein, T.M., E.D. Wolf, R. Wu, and J.C. Sanford. 1987. High velocity microprojectiles for delivering nucleic acids into living cells. Nature 327(6117):70–73.

Kosuri, S., and G.M. Church. 2014. Last-scale de novo DNA synthesis: Technologies and applications. Nature Methods 11:499–507.

Kotula, J.W., S.J. Kerns, L.A. Shaket, L. Siraj, J.J. Collins, J.C. Way, and P.A. Silver. 2014. Programmable bacteria detect and record an environmental signal in the mammalian gut. Proceedings of the National Academy of Sciences of the United States of America 111:4838–4843.

Kovalchuk, I., and O. Kovalchuk. 2008. Transgenic plants as sensors of environmental pollution genotoxicity. Sensors 8(3):1539–1558.

Kovalchuk, I., O. Kovalchuk, A. Arkhipov, and B. Hohn. 1998. Transgenic plants are sensitive bioindicators of nuclear pollution caused by the Chernobyl accident. Nature Biotechnology 16(11):1054–1059.

Kromdijk, J., K. Głowacka, L. Leonelli, S.T. Gabilly, M. Iwai, K.K. Niyogi, and S.P. Long. 2016. Improving photosynthesis and crop productivity by accelerating recovery from photoprotection. Science 354:857–861.

Kuiken, T. 2016. Our Collective Biology: Enabling Public Science to Build an Ecosystem of Makers in Biology. Presentation to the National Academies of Sciences, Engineering, and Medicine Committee on Future Biotechnology Products and Opportunities to Enhance Capabilities of the Biotechnology Regulatory System, June 1, Washington, DC.

Kwong, W.K., and N.A. Moran. 2016. Gut microbial communities of social bees. Nature Reviews Microbiology 14(6): 374–384.

Lafrance, A. September 24, 2015. Bioluminescence in a bottle. The Atlantic. Available at http://www.theatlantic.com/technology/archive/2015/09/bioluminescence-in-a-bottle/407152/. Accessed January 8, 2017.

Lajoie, M.J., A.J. Rovner, D.B. Goodman, H.-R. Aerni, A.D. Haimovich, G. Kuznetsov, J.A. Mercer, H.H. Wang, P.A. Carr, J.A. Mosberg, N. Rohland, P.G. Schultz, J.M. Jacobson, J. Rinehart, G.M. Church, and F.J. Isaacs. 2013. Genomically recoded organisms expand biological functions. Science 342(6156):357–360.

Langin, K. July 18, 2014. Genetic engineering to the rescue against invasive species? National Geographic. Available at http://news.nationalgeographic.com/news/2014/07/140717-gene-drives-invasive-species-insects-disease-science-environment/. Accessed January 8, 2017.

Ledford, H. 2016. UK bioethicists eye designer babies and CRISPR cows. Nature 538(7623):17.

Lee, H.H., N. Ostrov, B.G. Wong, M.A. Gold, A. Khalil, and G.M. Church. 2016. *Vibrio natriegens*, a new genomic powerhouse. bioRxiv:058487.

Li, F., Y. Li, H. Liu, H.H. Zhang, C.X. Liu, X.J. Zhang, H.W. Dou, W.X. Yang, and Y.T. Du. 2014. Production of GHR double-allelic knockout Bama pig by TALENs and handmade cloning. Hereditas 36(9):903–911.

Li, J., L. Gu, J. Aach, and G.M. Church. 2014. Improved cell-free RNA and protein synthesis system. PLOS ONE 9(9):e106232.

Liew, F., M.E. Martin, R.C. Tappel, B.D. Heijstra, C. Mihalcea, and M. Köpke. 2016. Gas fermentation—A flexible platform for commercial scale production of low-carbon-fuels and chemicals from waste and renewable feedstocks. Frontiers in Microbiology 7:694.

Lindemann, S.R., H.C. Bernstein, H.-S. Song, J.K. Fredrickson, M.W. Fields, W. Shou, D.R. Johnson, and A.S. Beliaev. 2016. Engineering microbial consortia for controllable outputs. The ISME Journal 10:2077–2084.

Lombardo, T. October 11, 2015. Bioluminescent lamp: Oddity, novelty, engineering challenge. Available at http://www.engineering.com/ElectronicsDesign/ElectronicsDesignArticles/ArticleID/10799/Bioluminescent-Lamp-Oddity-Novelty-Engineering-Challenge.aspx. Accessed January 8, 2017.

Luzar, C. August 3, 2013. Kickstarter bans GMOs in wake of glowing plant campaign. Crowdfund Insider. Available at http://www.crowdfundinsider.com/2013/08/20031-kickstarter-bans-gmos-in-wake-of-glowing-plant-fiasco. Accessed September 27, 2016.

Ma, N.J., and F.J. Isaacs. 2016. Genomic recoding broadly obstructs the propagation of horizontally transferred genetic elements. Cell Systems 3(2):199–207.

Mali, P., L. Yang, K.M. Esvelt, J. Aach, M. Guell, J.E. DiCarlo, J.E. Norville, and G.M. Church. 2013. RNA-guided human genome engineering via Cas9. Science 339(6121):823–826.

Mandell, D.J., M.J. Lajoie, M.T. Lee, R. Takeuchi, G. Kuznetsov, J.E. Norville, C.J. Gregg, B.L. Stoddard, and G.M. Church. 2015. Biocontainment of genetically modified organisms by synthetic protein design. Nature 518(7537):55–60.

Marcellin, F. February, 26, 2016. Glow-in-the-dark bacterial lights could illuminate shop windows. New Scientist. Available at https://www.newscientist.com/article/2078921-glow-in-the-dark-bacterial-lights-could-illuminate-shop-windows. Accessed January 8, 2017.

Marris, C., I.H. Langford, and T. O'Riordan. 1998. A quantitative test of the cultural theory of risk perceptions: Comparison with the psychometric paradigm. Risk Analysis 18(5):635–647.

McCright, A.M., and C. Xiao. 2014. Gender and environmental concern: Insights from recent work and for future research. Society & Natural Resources 27(10):1109–1113.

Miller, F. December 3, 2014. Arkansas releases first Roundup Ready soybean. University of Arkansas News. Available at http://arkansasagnews.uark.edu/8273.htm. Accessed November 27, 2016.

Mizokami, K. November 21, 2016. The Pentagon uses plant DNA to catch counterfeit parts. Popular Mechanics. Available at http://www.popularmechanics.com/military/research/a23988/plant-dna-pentagon-counterfeit/. Accessed January 8, 2017.

Munnelly, K. 2016. Advancements in DNA Construction Tools and Technology. Presentation to the National Academies of Sciences, Engineering, and Medicine Committee on Future Biotechnology Products and Opportunities to Enhance Capabilities of the Biotechnology Regulatory System, June 1, Washington, DC.

Nagare, P., B.A. Aglave, and M.O. Lokhande. 2009. Genetically engineered zebra fish—fluorescent beauties with practical applications. The Asian Journal of Animal Science 4(1):126–129.

Napolitano, M.G., M. Landon, C.J. Gregg, M.J. Lajoie, L. Govindarajan, J.A. Mosberg, G. Kuznetsov, D.B. Goodman, O. Vargas-Rodriguez, F.J. Isaacs, D. Söll, and G.M. Church. 2016. Emergent rules for codon choice elucidated by editing rare arginine codons in *Escherichia coli*. Proceedings of the National Academy of Sciences of the United States of America 113:E5588–E5597.

NASEM (National Academies of Sciences, Engineering, and Medicine). 2016a. Gene Drives on the Horizon: Advancing Science, Navigating Uncertainty, and Aligning Research with Public Values. Washington, DC: The National Academies Press.

NASEM. 2016b. Genetically Engineered Crops: Experiences and Prospects. Washington, DC: The National Academies Press.

OECD (Organisation for Economic Co-operation and Development). 2015. Building a Bioeconomy: How Nations Approach Capacity Building in Industrial Biotechnology. Paris: OECD.

OSTP (Office of Science and Technology Policy). 2012. National Bioeconomy Blueprint. Washington, DC: The White House.

Ostrov, N., M. Landon, M. Guell, G. Kuznetsov, J. Teramoto, N. Cervantes, M. Zhou, K. Singh, M.G. Napolitano, and M. Moosburner. 2016. Design, synthesis, and testing toward a 57-codon genome. Science 353(6301):819–822.

Oxitec. 2016. Draft Environment Assessment for Investigational Use of *Aedes aegypti* OX513A. February. Oxfordshire, UK: Oxitec, Ltd. Available at http://www.fda.gov/downloads/AnimalVeterinary/DevelopmentApprovalProcess/Genetic Engineering/GeneticallyEngineeredAnimals/UCM487377.pdf. Accessed September 30, 2016.

Peck, B. 2016. Reimagine SequenceSpace. Presentation to the National Academies of Sciences, Engineering, and Medicine Committee on Future Biotechnology Products and Opportunities to Enhance Capabilities of the Biotechnology Regulatory System, June 1, Washington, DC.

Petzold, C.J., L.J.G. Chan, M. Nhan, and P.D. Adams. 2015. Analytics for metabolic engineering. Frontiers in Bioengineering and Biotechnology 3:135.

Potter, K.M., and B.L. Conkling, eds. 2016. Forest Health Monitoring: National Status, Trends, and Analysis 2015. Asheville, NC: U.S. Forest Service–Southern Research Station.

Radakovits, R., R.E. Jinkerson, A. Darzins, and M.C. Posewitz. 2010. Genetic engineering of algae for enhanced biofuel production. Eukaryotic Cell 9(4):486–501.

Reardon, S. 2016. Welcome to the CRISPR zoo. Nature 431(7593):160–163.

Reed, T. 2016. Intrexon. Presentation to the National Academies of Sciences, Engineering, and Medicine Committee on Future Biotechnology Products and Opportunities to Enhance Capabilities of the Biotechnology Regulatory System, June 1, Washington, DC.

Regalado, A. July 15, 2016. Why Kickstarter's glowing plant left backers in the dark. MIT Technology Review. Available at https://www.technologyreview.com/s/601884/why-kickstarters-glowing-plant-left-backers-in-the-dark. Accessed September 27, 2016.

Roberts, J.P., S. Stauffer, C. Cummings, and J. Kuzma. 2015. Synthetic Biology Governance: Delphi Study Workshop Report. GES Center Report No. 2015.2. Available at https://research.ncsu.edu/ges/files/2014/04/Sloan-Workshop-Report-final-ss-081315-1.pdf. Accessed October 10, 2016.

Rohrmann, B., and O. Renn. 2000. Risk perception research. Pp. 11–53 in Cross-Cultural Risk Perception: A Survey of Empirical Studies, O. Renn and B. Rohrmann, eds. New York: Springer.

Rovner, A.J., A.D. Haimovich, S.R. Katz, Z. Li, M.W. Grome, B.M. Gassaway, M. Amiram, J.R. Patel, R.R. Gallagher, J. Rinehart, and F.J. Isaacs. 2015. Recoded organisms engineered to depend on synthetic amino acids. Nature 518(7537):89–93.

Samin, G., M. Pavlova, M.I. Arif, C.P. Postema, J. Damborsky, and D.B. Janssen. 2014. A *Pseudomonas putida* strain genetically engineered for 1,2,3-trichloropropane bioremediation. Applied and Environmental Microbiology 80(17): 5467–5476.

San Miguel, K., and J.G. Scott. 2016. The next generation of insecticides: dsRNA is stable as a foliar-applied insecticide. Pest Management Science 72(4):801–809.

Schippers, A., S. Hedrich, J. Vasters, M. Drobe, W. Sand, and S. Willscher. 2014. Biomining: Metal recovery from ores with microorganisms. Advances in Biochemical Engineering/Biotechnology 141:1–47.

SEC (U.S. Securities and Exchange Commission). 2016. Investor Bulletin: Crowdfunding for Investors. February 16. Available at https://www.sec.gov/oiea/investor-alerts-bulletins/ib_crowdfunding-.html. Accessed September 27, 2016.

Segran, E. November 4, 2015. Twist Bioscience inks deal to sell 100 million base pairs of synthetic DNA. Fast Company. Online. Available at https://www.fastcompany.com/3053065/most-creative-people/twist-bioscience-inks-deal-to-sell-100-million-pairs-of-synthetic-dna. Accessed October 11, 2016.

Seltenrich, N. July 23, 2015. Improving the work of silkworms and spiders, with yeast. Berkeley Engineering. Available at http://engineering.berkeley.edu/2015/07/improving-work-silkworms-and-spiders-yeast. Accessed January 8, 2017.

Sewalt, V. 2016. Managing Risks Associated with Biotechnology Used in Containment. Presentation to the National Academies of Sciences, Engineering, and Medicine Committee on Future Biotechnology Products and Opportunities to Enhance Capabilities of the Biotechnology Regulatory System, June 27, San Francisco, CA.

Shapiro, B. 2015. How to Clone a Mammoth: The Science of De-Extinction. Princeton, NJ: Princeton University Press.

Shi, J., H. Gao, H. Wang, H.R. Lafitte, R.L. Archibald, M. Yang, S.M. Hakimi, H. Mo, and J.E. Habben. 2017. ARGOS8 variants generated by CRISPR-Cas9 improve maize grain yield under field drought stress conditions. Plant Biotechnology Journal 15(2):207–216.

Shigeta, R. 2016. Indie Bio: New Opportunities, a New Generation of Biotech Entrepreneurs. Presentation to the National Academies of Sciences, Engineering, and Medicine Committee on Future Biotechnology Products and Opportunities to Enhance Capabilities of the Biotechnology Regulatory System, June 27, San Francisco, CA.

Shiva, V., D. Barker, and C. Lockhart. 2011. The GMO Emperor Has No Clothes: A Global Citizens Report on the State of GMOs, Synthesis Report. Florence: Navdanya International.

Shontell, A. June 28, 2016. A Brooklyn startup that's armed with $40 million is growing real leather in a lab without hurting a single animal. Business Insider. Available at http://www.businessinsider.com/modern-meadow-lab-grown-leather-2016-6. Accessed January 8, 2017.

Slovic, P. 1987. Perception of risk. Science 236:280–290.

Smanski, M.J., S. Bhatia, D. Zhao, Y. Park, L.B.A. Woodruff, G. Giannoukos, D. Ciulla, M. Busby, J. Calderon, R. Nicol, D.B. Gordon, D. Densmore, and C.A. Voigt. 2014. Functional optimization of gene clusters by combinatorial design and assembly. Nature Biotechnology 32:1241–1249.

Stanton, B. 2016. Ginkgo Bioworks. Presentation to the National Academies of Sciences, Engineering, and Medicine Committee on Future Biotechnology Products and Opportunities to Enhance Capabilities of the Biotechnology Regulatory System, June 1, Washington, DC.

Sun, Z.Z., V. Noireaux, and R.M. Murray. 2014. Accelerating the design-build-test cycle of synthetic biological circuits in E. coli using S30 TX-TL cell-free systems, linear DNA, and modular assembly. General submission in the Conference Proceedings of the Synthetic Biology: Engineering, Evolution & Design Conference, July 14–17, Manhattan Beach, CA. Available at http://www3.aiche.org/proceedings/Abstract.aspx?PaperID=392699. Accessed September 27, 2016.

Taylor, N., E. Gaitán-Solís, T. Moll, B. Trauterman, T. Jones, A. Pranjal, C. Trembley, V. Abernathy, D. Corbin, and C. Fauquet. 2012. A high-throughput platform for the production and analysis of transgenic cassava (Manihot esculenta) plants. Tropical Plant Biology 5(1):127–139.

Tennessen, J.A., A. Théron, M. Marine, J.-Y. Yeh, A. Rognon, and M.S. Blouin. 2015. Hyperdiverse gene cluster in snail host conveys resistance to human schistosome parasites. PLOS Genetics 11(3):e1005067.

Thomke, S.H. 2003. Experimentation Matters: Unlocking the Potential of New Technologies for Innovation. Boston: Harvard Business Press.

Thompson, P. 2011. Understanding and coping with social risk in emerging technology risk assessment. Pp. 1–16 in Biotechnology and Nanotechnology Risk Assessment: Minding and Managing the Potential Threats around Us, S. Ripp and T.B. Henry, eds. Washington, DC: American Chemical Society.

Tokareva, O., V.A. Michalczechen-Lacerda, E.L. Rech, and D.L. Kaplan. 2013. Recombinant DNA production of spider silk proteins. Microbial Biotechnology 6(6):651–663.

Trögl, J., A. Chauhan, S. Ripp, A.C. Layton, G. Kuncová, and G.S. Sayler. 2012. Pseudomonas fluorescens HK44: Lessons learned from a model whole-cell bioreporter with a broad application history. Sensors 12:1544–1571.

USDA–APHIS (U.S. Department of Agriculture–Animal and Plant Health Inspection Service). 2017. Regulatory Impact Analysis & Initial Regulatory Flexibility Analysis, Proposed Rule, APHIS 2015-0057, RIN 0579-AE15; Importation, Interstate Movement, and Environmental Release of Organisms Produced through Genetic Engineering (7 CFR part 340). Available at https://www.regulations.gov/document?D=APHIS-2015-0057-0002. Accessed March 31, 2017.

USDA–NASS (U.S. Department of Agriculture–National Agricultural Statistics Service). 2009. Hawaii Papayas. Available at https://www.nass.usda.gov/Statistics_by_State/Hawaii/Publications/Fruits_and_Nuts/papaya.pdf. Accessed December 3, 2016.

Van Beilen, J.B., and Y. Poirier. 2008. Production of renewable polymers from crop plants. The Plant Journal 54(4):684–701.

van der Linden, S. 2016. A conceptual critique of the cultural cognition thesis. Science Communication 38(1):128–138.

Vidal, J. October 19, 2011. GM foods: A "biotech revolution"? The Guardian. Available at https://www.theguardian.com/environment/2011/oct/19/gm-foods-a-biotech-revolution. Accessed January 30, 2017.

Voytas, D.F., and C. Gao. 2014. Precision genome engineering and agriculture: Opportunities and regulatory challenges. PLOS Biology 12:e1001877.

Waltz, E. 2016. Gene-edited CRISPR mushroom escapes US regulation. Nature 532(7599):293.

Wang, H.H., F. J. Isaacs, P.A. Carr, Z.Z. Sun, G. Xu, C.R. Forest, and G.M. Church. 2009. Programming cells by multiplex genome engineering and accelerated evolution. Nature 460(7257):894–898.

Wang, Y., X. Cheng, Q. Shan, Y. Zhang, J. Liu, C. Gao, and J.L. Qiu. 2014. Simultaneous editing of three homoeoalleles in hexaploid bread wheat confers heritable resistance to powdery mildew. Nature Biotechnology 32:947–951.

Way, J.C., J.J. Collins, J.D. Keasling, and P.A. Silver. 2014. Integrating biological redesign: Where synthetic biology came from and where it needs to go. Cell 157(1):151–161.

Weinstock, M.T., E.D. Hesek, C.M. Wilson, and D.G. Gibson. 2016. Vibrio natriegens as a fast-growing host for molecular biology. Nature Methods 13:849–851.

Wheelwright, S.C., and K.B. Clark. 1994. Accelerating the design-build-test cycle for effective product development. International Marketing Review 11:32–46.

Wozniak, C.A., G. McClung, J. Gagliardi, M. Segal, and K. Matthews. 2012. Regulation of genetically engineered microorganisms under FIFRA, FFDCA and TSCA. Pp. 57–94 in Regulation of Agricultural Biotechnology: The United States and Canada, C. Wozniak and A. McHughen, eds. Heidelberg, Germany: Springer.

Xiong, W., J.A. Morgan, J. Ungerer, B. Wang, P.-C. Maness, and J. Yu. 2015. The plasticity of cyanobacterial metabolism supports direct CO_2 conversion to ethylene. Nature Plants 1:15053.

Xu, T., D.M. Close, G.S. Sayler, and S. Ripp. 2013. Genetically modified whole-cell bioreporters for environmental assessment. Ecological Indicators 28:125–141.

Yue, C., S. Zhao, C. Cummings, and J. Kuzma. 2015. Investigating factors influencing consumer willingness to buy GM food and nano-food. Journal of Nanoparticle Research 17(7):283.

Zarate, X., and D.W. Galbraith. 2014. A cell-free expression platform for production of protein microarrays. Methods in Molecular Biology 1118:297–307.

Zetsche, B., J.S. Gootenberg, O.O. Abudayyeh, I.M. Slaymaker, K.S. Makarova, P. Essletzbichler, S.E. Volz, J. Joung, J. van der Oost, A. Regev, and E.V. Koonin. 2015. Cpf1 is a single RNA-guided endonuclease of a class 2 CRISPR-Cas system. Cell 163(3):759–771.

Zhang, B., A.D. Oakes, A.E. Newhouse, K.M. Baier, C.A. Maynard, and W.A. Powell. 2013. A threshold level of oxalate oxidase transgene expression reduces *Cryphonectria parasitica*-induced necrosis in a transgenic American chestnut (*Castanea dentata*) leaf bioassay. Transgenic Research 22(5):973–982.

Zhang, L., R. Routsong, Q. Nguyen, E.L. Rylott, N.C. Bruce, and S.E. Strand. 2017. Expression in grasses of multiple transgenes for degradation of munitions compounds on live fire training ranges. Plant Biotechnology Journal 15(5):624–633.

Zhou, H., B. Vonk, J.A. Roubos, R.A.L. Bovenberg, and C.A. Voigt. 2015. Algorithmic co-optimization of genetic constructs and growth conditions: Application to 6-ACA, a potential nylon-6 precursor. Nucleic Acids Research 43(2):10560–10570.

Zhu, H., L. Hu, J. Liu, H. Chen, C. Cui, Y. Song, Y. Jin, and Y. Zhang. 2016. Generation of β-lactoglobulin-modified transgenic goats by homologous recombination. FEBS Journal 283(24):4600–4613.

Zou, Q., X. Wang, Y. Liu, Z. Ouyang, H. Long, S. Wei, J. Xin, B. Zhao, S. Lai, J. Shen, Q. Ni, H. Yang, H. Zhong, L. Li, M. Hu, Q. Zhang, Z. Zhou, J. He, Q. Yan, N. Fan, Y. Zhao, Z. Liu, L. Guo, J. Huang, G. Zhang, J. Ying, L. Lai, and X. Gao. 2015. Generation of gene-target dogs using CRISPR/Cas9 system. Journal of Molecular Cell Biology 7(6):580–583.

3

The Current Biotechnology Regulatory System

This chapter introduces the existing risk-analysis system for biotechnology products, surveying agency authorities as they relate to future biotechnology products anticipated over the next 5–10 years. Consistent with this study's statement of task, the focus is on the U.S. Environmental Protection Agency (EPA), the U.S. Food and Drug Administration (FDA), and the U.S. Department of Agriculture (USDA) but not limited to these agencies. This discussion highlights major statutes that authorize these agencies either to conduct risk analysis or to require other entities—such as product developers—to conduct it.

The regulatory agencies carry out two closely interrelated but distinct regulatory functions that together protect public health, safety, and the environment and which form the major topics of this chapter. The first is consumer- and occupational-safety regulations that protect members of the public who directly expose themselves to biotechnology products through their decisions to consume or use them or to enter a workplace where biotechnology products or biotechnological means of production are in use. Examples of possible risks associated with these products are injuries consumers may suffer when using a biotechnology-based product or injuries to industrial workers caused by exposure to a biotechnological means of production. The other function is environmental regulation to address non–human health risks (that is, ecological risks) and human health risks to members of the public that are exposed to future biotechnology products regardless of their individual decisions. Examples of these possible risks include contamination of the surrounding environment and introduction of a pernicious species.

The broad, environmental impacts of biotechnology products have historically been a focus of concern and this can be expected to continue in the future. The expanding array of products identified in Chapter 2 suggests, however, that consumer- and occupational-safety issues may assume growing importance within the time frame of this study. The 2017 update to the Coordinated Framework (EOP, 2017) devotes less attention to these issues than to environmental risks, so the committee opted to examine consumer- and occupational-safety issues in depth in this chapter before introducing the key statutes related to environmental risks, which lay groundwork for more detailed discussion of environmental risk analysis that follows in Chapters 4 and 5 of this report.

The temporal limitation of this study's scope necessarily implies a focus on legal authorities conferred by existing federal statutes. Major shifts in federal agencies' approach to risk analysis often evolve over a time scale longer than 5–10 years. Even after Congress enacts a new statute—a process that itself takes time—the process of implementing a new statute may include legal challenges and requires agencies to promulgate implementing regulations. Because regulatory change is a long process, some of the new biotechnology products described in Chapter 2 will likely challenge federal agencies to protect the public's health, welfare, safety, and environment using the legal tools they already have. This chapter describes key features of the risk-analysis frameworks provided by statutes enacted at the time the committee was writing its report, including the 2016 amendments[1] to the Toxic Substances Control Act (TSCA),[2] and touches on scientific capabilities, tools, and expertise that may be useful to the agencies as they face this challenge; these matters are discussed in more detail in Chapter 4. The broader task of identifying potentially beneficial statutory amendments was outside the scope of this study, but this chapter does, when appropriate, highlight some of the costs, limitations, gaps, and redundancies that may arise as agencies attempt to regulate future biotechnology products using the legal authorities conferred by their existing statutes.

In January 2017, the federal government released an update to the Coordinated Framework (EOP, 2017), which provided a detailed overview of the responsibilities of EPA, FDA, and USDA within the Coordinated Framework and the major statutes these agencies enforce. Consequently, this chapter directs its attention to aspects of these statutes that are central to the report's statement of task. One important question concerns the breadth of jurisdiction conferred by these statutes: Will all of the new biotechnology products expected over the next 5–10 years fit within the categories of products that EPA, FDA, and USDA can regulate? If not, which (if any) federal agencies will regulate the product? What authorities does each responsible agency have to conduct (or require) risk analysis, and what scientific capabilities, tools, and expertise may be useful to the agencies as they confront the new biotechnology products identified in Chapter 2? Finally, how much flexibility do the agencies have under their existing statutes to address any gaps that threaten to leave new products with inadequate regulatory oversight?

After a brief overview of the origins and evolution of the Coordinated Framework, the chapter addresses the statutory authorities and associated agency responsibilities for accomplishing the Coordinated Framework's objectives with regard to safety: protection of human health—including consumer and occupational safety—and protection of the environment. It identifies the capacity of the agencies to regulate future biotechnology products under their existing authorities and points out areas where the limits of their authorities may leave gaps in oversight.

OVERVIEW OF U.S. REGULATORY SYSTEM

The purpose of the U.S. regulatory system is to "protect public health, welfare, safety, and our environment while promoting economic growth, innovation, competitiveness, and job creation" (EOP, 2011:3821). To strike this balance, *risk* in the U.S. system has been understood to pertain to consumer safety and environmental protection; however, the term can be defined more broadly (Box 3-1). This section provides a brief history and evolution of the Coordinated Framework, which was established to provide oversight for risks to human health and the environment. It also reviews, in more general terms, the considerations that go into balancing safety through regulations with innovation.

---

[1]Frank R. Lautenberg Chemical Safety for the 21st Century Act, P.L. 114-182 (2016).
[2]15 U.S.C. § 2601 *et seq.*

BOX 3-1
Conceptions of Risk

Kaplan and Garrick (1981) defined a *risk scenario* as comprising three questions—what can happen, how likely is that to happen, and if it does happen what are the consequences. They also emphasized the importance of examining all possible ways that things may go wrong. Renn (1992) observed that the term *risk* contained three elements: undesirable outcomes (what), possibility of occurrence (how likely), and state of reality (ways it occurs in pathways), which complement the three questions from Kaplan and Garrick. Individuals may weigh these elements differently, which leads to a broad range of perspectives, some of which are oriented more toward technical or economic risks and others toward social or cultural risks. Similarly, frameworks for analyzing risk vary from the linear assessment process for direct human health and environmental harms (NRC, 1983) to the more iterative and engaged processes that consider various types of harms, including socioeconomic effects and cultural affronts and pathways through which they manifest risk (Renn, 2005; Stirling, 2007; IRGC, 2015).

In the United States, regulatory *risk assessment* is largely confined to human, ecological, and economic conceptions of risk. Human health and ecological risk assessments concentrate on identifying (a) possible causes of harm (including the strength of the evidence of causation), (b) the relationship between exposure to the harm and the probability of the adverse effect, (c) the extent of human or environmental exposure to the harm, and (d) the probability of the harm occurring and the magnitude of the possible harm, taking into account bounds of uncertainty (NRC, 1983). However, people see many types of possible harms from new products to their well-being and way of life beyond those to human health and the environment. These can include the loss of employment (which can occur even in the situation where net economic benefits to society are positive), changes in social structures or relationships, cultural affronts, loss of biodiversity and the intrinsic value it provides, changing landscapes, shifts in power and privilege, growing inequities, and indirect or secondary harms on systems or landscape scales. Risk–cost–benefit analyses set the bounds of both the ethical and scientific judgments that may be considered when making a regulatory decision (Fischhoff, 2015).

Values are always embedded in risk analysis by the choices and interpretations of the people conducting them and the selection of risk-assessment endpoints of concern, methods, and questions (for example, Shrader-Frechette, 2007; Thompson, 2007; Kuzma and Besley, 2008). *Risk analysis* includes not just *risk assessment* but also *risk management*—that is, "the process of weighing policy alternatives and selecting the most appropriate regulatory action [after] integrating the results of risk assessment with engineering data and social, economic, and political concerns" (NRC, 1983:3).

Despite the diversity of risks that people care about and the knowledge they bring to the table about their own social structures, environments, and exposure pathways, risk analysis in the U.S. regulatory process has often been limited to industry developers, government regulators, and sometimes external advisory committees (Meghani and Kuzma, 2011); however, in some cases, EPA, FDA, and USDA have used public participation and external peer review as general practice (see Chapter 4). It has been suggested by scholars and think-tanks that more engaged and iterative risk-analysis frameworks could be designed to incorporate parameters and values important to stakeholders (NRC, 1996, 2009; Renn, 2005; IRGC, 2015). These approaches also address the emerging "risk society" in which globalization, complex and embedded technologies, and abrupt events combine to increase the pace, scale, and spread of risks, making them embedded, ubiquitous, and central to societies (Rosa et al., 2013). A new risk governance paradigm goes beyond traditional risk analysis and includes a more open, iterative, and engaged design (Rosa et al., 2013).

The Coordinated Framework

Federal involvement in the oversight of biotechnology is generally viewed as originating in the 1970s. Responding to concerns raised by scientists engaged in recombinant DNA research, the National Institutes of Health (NIH) published a set of research guidelines in 1976, which have been updated many times over the years (NIH, 2016). The NIH guidelines represent a formal research governance approach that applies to all research with recombinant or synthetic nucleic acid molecules conducted or sponsored by public and private institutions that receive any NIH funding for such research, and many other federal agencies and private research sponsors also require compliance with the NIH guidelines in research that they fund.[3] The NIH guidelines govern standards for protection of researchers, the public, and the environment.[4] NIH-funded research at institutions in the United States and abroad that conduct basic and clinical research involving recombinant or synthetic nucleic acid molecules must adhere to safety practices and containment procedures described in the NIH guidelines through the oversight of Institutional Biosafety Committees.[5] Even though the NIH guidelines may not govern all government-funded and privately funded research, they are a tool for the entire research community to understand the potential biosafety implications of recombinant or synthetic nucleic acid research.

Later in the 1970s and early 1980s, amid growing prospects that DNA research would produce a flood of new products, Congress considered but did not adopt various proposals to enact unified biotechnology legislation. In 1984, the White House Cabinet Council on Natural Resources and the Environment formed a Working Group on Biotechnology, which proposed a Coordinated Framework that would clarify regulatory responsibility to federal agencies acting under their existing statutory authorities (OSTP, 1986).

The Coordinated Framework was published in July 1986 after an 18-month period for public comment (OSTP, 1986). It orchestrated the biotechnology-related responsibilities of multiple federal agencies, with prominent roles assigned to NIH, EPA, FDA, USDA, and the U.S. Department of Labor's Occupational Safety and Health Administration (OSHA). The biotechnology-related responsibilities of each agency were conferred by statutes already in effect as of 1986, meaning that statutes enacted before the biotechnology revolution were interpreted to cover biotechnology. For example, "chemicals" under TSCA were interpreted as including biotechnology products (described in more detail below) (EPA, 1997).

In addition to regulating biotechnology as required by their respective enabling statutes, these agencies also comply with generally applicable federal statutes including the National Environmental Policy Act (NEPA), the Endangered Species Act (ESA), and the Administrative Procedure Act (APA). These latter statutes, while often characterized as procedural in nature, require agencies to conduct certain risk-analysis activities or, in the case of APA, facilitate public deliberation by subjecting agency activities to transparency and due-process requirements.

The Coordinated Framework was updated in 1992 to provide further policy guidance to agencies. That guidance stated that products intended for use in the environment should not be regulated on the basis of the process by which they were made; instead, the criteria would be "characteristics of the organism, the target environment, and the type of application" (OSTP, 1992:6755). In January 2017, the federal government published an update to the Coordinated Framework, the first in more than 20 years (EOP, 2017). That update provided a basic survey of statutory authorities of EPA, FDA, and USDA to regulate environmental and human health and safety risks related to biotechnology products (EOP, 2017). At the time a proposed update to the Coordinated Framework was put forward for public comment (September 2016)—before publishing the final update in January

---

[3]See "Product-Development Research" below for discussion of the impacts that anticipated shifts in the funding of biotechnology research may have on the continued applicability of the NIH guidelines.

[4]NIH Office of Science Policy. Biosafety. Available at http://osp.od.nih.gov/office-biotechnology-activities/biosafety. Accessed December 6, 2016.

[5]The responsibilities of the Institution Biosafety Committees extend to *research* involving material transfers among laboratories and field releases of genetically engineered organisms to governance of emerging technology applications, such as genome editing, or the design of gene-drive experiments and containment plans specific for research with gene drives. However, their responsibilities do not extend to biotechnology *products*.

2017—the federal government also published the *National Strategy for Modernizing the Regulatory System for Biotechnology* (EOP, 2016).

Table 3-1, originally published in the 2017 update to the Coordinated Framework, summarizes the statutes under which EPA, FDA, and USDA currently regulate biotechnology products. Under these statutes, the Coordinated Framework agencies carry out the two interrelated but distinct regulatory functions described earlier: (1) consumer- and occupational-safety regulation and (2) environmental regulation.

After considering the way regulations take into account safety and innovation, this chapter delves into the following two topics: (1) whether the statutes in Table 3-1 provide the agencies with adequate tools to support robust risk assessment to protect human health and the environment and (2) whether the agencies have sufficient statutory powers to address the special challenges raised by new products expected during the next 5–10 years.

Regulation to Promote Safety and Innovation

When the Coordinated Framework was established, one of its stated purposes was to find a balance between safety regulation and innovation (OSTP, 1986; Box 3-2). The two are not necessarily at odds: innovation has the potential to enhance safety, for example, by replacing high-risk products with newer, safer products. Moreover, regulatory frameworks that are suitably designed and implemented have the potential to foster innovation; for example, fuel economy standards have improved the average fuel economy of U.S. vehicles. Thus, it is incorrect to assume that regulation inevitably creates barriers to innovation. One valid area of concern, however, is that regulations have the potential to impede or delay the introduction of innovative products to the market, if the regulations add substantial up-front costs and delays to the process of developing and marketing a new product. These costs and delays can also provide incentives for developers to create products that are not considered regulated articles (such as in the example of using biolistics to transform a grass species described in Chapter 2). A related concern is that regulation can encourage developers to imitate products that have charted a path through the regulatory system, rather than pursue more innovative products that may have unclear paths and thus run the risk of taking a long time in regulatory review. The intent of the Coordinated Framework is to provide mechanisms to assess the safety of biotechnology products while simultaneously offering a framework for advancing innovation and increasing transparency, coordination, efficiency, and predictability. This balance is sought in the interpretation of the statutory authorities that make up the Coordinated Framework.

It is simplistic to characterize an entire statute as either precautionary or conducive to innovation, as these concepts have meaning only at the level of specific statutory provisions. Consumer- and occupational-safety statutes generally address risks through a complex mix of provisions that include some precautionary and some permissive elements. Together, a statute's provisions balance the need to be cautious against the benefits people may gain as innovative new products enter the marketplace. Among the features that are important in accessing the risk characteristics of a specific statutory framework are

- The allocation of duties to develop an information base to support regulatory decision making.
- Where the burden of proof is placed in regulatory decisions.
- The mix of premarket and post-market safety information.
- The provisions that foster private-sector participation in data and evidence generation for risk analysis.
- The provisions for managing risks that are revealed during a risk assessment.

TABLE 3-1 Statutes and Protection Goals Related to the U.S. Environmental Protection Agency (EPA), the U.S. Food and Drug Administration (FDA), and the U.S. Department of Agriculture (USDA) for the Regulation of Biotechnology Products

Agency	Statute	Protection Goal
EPA	Federal Insecticide, Fungicide, and Rodenticide Act (FIFRA)	Prevent and eliminate unreasonable adverse effects on the environment • For environmental and occupational risks, this involves comparing economic, social, and environmental risks to human health and the environment and benefits associated with the pesticide use. • For dietary or residential human health effects, the sole standard is the "safety" of all the combined exposures to the pesticide and related compounds.
EPA	Federal Food, Drug, and Cosmetic Act (FDCA)	Ensure that no harm will result from aggregate exposure to the pesticide chemical residue, including all anticipated dietary exposures and all other exposures for which there is reliable information.
EPA	Toxic Substances Control Act (TSCA)	Prevent the manufacture, processing, distribution in commerce, use, or disposal of chemical substances, or any combination of such activities with such substances, from presenting an unreasonable risk of injury to health or the environment, including an unreasonable risk to a potentially exposed or susceptible population, without consideration of costs or other nonrisk factors.
FDA	FDCA	Ensure human and animal food is safe, sanitary, and properly labeled. Ensure human and animal drugs are safe and effective. Ensure the reasonable assurance of the safety and effectiveness of devices intended for human use. Ensure cosmetics are safe and properly labeled.
FDA	Public Health Service Act	Ensure the safety, purity, and potency of biological products.
USDA	Animal Health Protection Act (AHPA)	Protect livestock from animal pest and disease risks.
USDA	Plant Protection Act (PPA)	Protect agricultural plants and agriculturally important natural resources from damage caused by organisms that pose plant pest or noxious weed risks.
USDA	Federal Meat Inspection Act	Ensure that the United States' commercial supply of meat, poultry, and egg products is safe, wholesome, and correctly labeled.
USDA	Poultry Products Inspection Act	Ensure that the United States' commercial supply of meat, poultry, and egg products is safe, wholesome, and correctly labeled.
USDA	Egg Products Inspection Act	
USDA	Virus-Serum-Toxin Act	Ensure that veterinary biologics are pure, safe, potent, and effective.

SOURCE: EOP (2017:9).

BOX 3-2
Risk Assessment and Use Restrictions

The level of risk a product presents may depend on how it is sold, distributed, or used. Products that are otherwise safe may pose high risks in the hands of unqualified or malicious users. For example, the safety and biosecurity of do-it-yourself biology (DIYbio) products may be determined by the skill or the intent of the user. In the area of human health, a drug that treats a legitimate medical condition may pose a high risk if used inappropriately, but FDA's power to limit human drug and device products to prescription use helps manage safety risks and enhances risk assessment by appointing a learned intermediary—the clinician—to perform a patient-specific risk assessment after considering the risks and benefits in light of the patient's characteristics. However, for the nondrug and nonmedical device products covered in the committee's report, federal consumer-safety statutes administered by FDA and the Consumer Product Safety Commission (CPSC) do not empower these regulators to restrict who is allowed to receive a product. EPA has some authority to restrict access and use of pesticides under the Federal Insecticide, Fungicide, and Rodenticide Act (FIFRA).[a] Under the Toxic Substances Control Act (TSCA), EPA sets conditions on the approval of a premarket notification or a microbial commercial activity notification to ensure safety to human health and the environment, for example, by establishing requirements for containment in the manufacture of the products or for personal protective equipment. EPA's authority under TSCA also allows it to require (through a consent order) that downstream users of a product use specified risk-mitigation measures; however, at the time the committee was writing its report, the means to enforce such an order was unclear. Therefore, in the example of the DIYbio product, regulators may not be able conduct a risk assessment if the product falls outside the statutory authorities. More generally, existing consumer-safety statutes provide only limited power—or in most cases no power at all—for regulators to restrict use, sale, and distribution so as to keep biotechnology products out of the hands of people whose use of the product could be dangerous either for themselves or others.

There are various federal programs that restrict access to dangerous biological agents. For example, the Federal Select Agent Program administered by the U.S. Department of Agriculture's Animal and Plant Health Inspection Service and the Centers for Disease Control and Prevention tracks the use of approximately 65 highly dangerous biological agents (for example, Ebola virus), and the U.S. Department of Commerce restricts shipments of certain chemicals and may require a license for some transactions (Lin, 2013). There is also voluntary screening of synthetic DNA orders by DNA providers, and the U.S. Department of Health and Human Services encourages participation (Lin, 2013). The Federal Bureau of Investigation's Biological Countermeasures Unit is active in identifying and responding to threats related to DIYbio with a focus on the potential for bioterrorism (You, 2016). These programs, however, are neither focused nor scaled to address the risks of diverse biotechnology consumer products expected in coming years, and existing consumer-safety regulators like FDA, CPSC, and EPA lack statutory tools to take on this responsibility.

---

[a]Under FIFRA, EPA can register the product as a general use product if it is deemed to be safe for users following general ("simple to understand") use instructions. EPA can register the product as a restricted use pesticide if it can be used safely only by—or under the supervision of—a person who has a pesticide applicator training certification (a program implemented by the states that typically applies to some agricultural pesticides, but not to household/residential use pesticides). Restricted use pesticides can only be lawfully sold to a person (entity) that has an applicator certification. It also is unlawful for someone without a certification, or not under the direct supervision of someone with a certificate, to use a restricted use pesticide. If EPA determines a pesticide cannot be used safely by a certified applicator with appropriate personal protection equipment, the agency declines to register the new product (or specific uses for the product) or will modify (i.e., delete the use or modify the use) or cancel an existing product's registration.

Allocation of Duties to Develop an Information Base to Support Regulatory Decision Making

In order to make sound regulatory decisions that protect the public and the environment, regulators need a base of reliable information about the risks a product may pose. Statutes provide various mechanisms to facilitate creation of this information base. First, the regulatory agency (or group of agencies) needs to know that the product exists; then, the regulatory agency needs information about the product in order to make a regulatory decision.

With regard to learning about products still in development, one option is for the statute to require product sponsors[6] to make their activities and new products known to the regulator via a notification or other registration requirement so as to facilitate rapid detection and response if safety problems later arise. If there is not a statutory requirement for such notification, then the regulator must instead conduct market surveillance to discover the existence of new products and the identities of their manufacturers.

Statutes also allocate responsibilities surrounding generation of safety information, by identifying who is responsible for conducting studies to assess risks and resolve the scientific uncertainties that can surround regulatory decisions involving novel biotechnology products. A statute may foster safety studies by the product sponsors (manufacturers, distributors, or sellers) who wish to benefit commercially from the new product and provide for investigational use of unapproved products under regulatory oversight. It may also require product sponsors to conduct (and fund) safety studies and require sponsors to submit data at key points during the product life cycle, such as first open release in the environment, initial market entry, or after consumer-safety incidents occur. Alternatively, a statute (or lack thereof) may require the regulatory agency to conduct its own studies (or rely on publicly funded research) to evaluate product safety. The nature of the statute dictates the distribution of responsibilities for generating safety data between the private product sponsor and federal agencies.

Requiring product sponsors to bear heavy evidence-generation burdens in the premarket period can raise the barriers to entry and the cost of bringing new products on to the market and thus may deter innovation, especially if the projected future market value of the product is estimated to be less than the combined costs of discovery, product development, and safety evaluation. Yet regulators and public-research funding agencies may lack resources to finance all the risk assessment that is needed to evaluate the public's safety in a time of fast-paced introduction of novel products. How statutes allocate responsibilities to generate evidence between private and public actors is therefore a critical parameter affecting the balance between innovation and safety.

Placement of the Burden of Proof for Regulatory Decision Making

A closely related question is where a statute places the burden of proof in key regulatory decisions, such as the decision to allow commercial sale of a new product or to add a new safety warning or restrict sales of an existing one. Faced with scientific uncertainty and evolving safety information, a regulator's ability to manage safety risks may hinge on where the burden of proof is placed and on the implicit presumptions embedded in the law: is the product presumed dangerous until proved safe, or is it treated as safe until proved dangerous (Charo, 2015)? When data are insufficient to prove either danger or safety, where that presumption falls often is determinative (Charo, 2015).

Where this burden of proof is placed also affects the cost of new product development and, therefore, may affect the pace of innovation. When the regulator bears the burden of proving products to be unsafe, this relieves product developers of the cost of proving their products safe, but it implies that the regulator's budget (or governmentally funded research) must bear primary responsibility for ensuring consumer, occupational, and environmental safety. Federal budgetary constraints may at times make it difficult to generate all of the data necessary to promote optimal levels of safety. Enacting statutes that require product developers to prove their products safe shifts

---

[6]In its discussion of FDA's authority, the committee has used the term *sponsor* to be consistent with language in the statutory authorities that pertain to the agency. The committee uses this term as it has used product *developer* elsewhere in the report, which can refer to a person, corporation, or agency that has brought a new product to a regulatory agency for oversight or regulatory approval.

this cost to them, but may also make products more costly if developers add research costs into pricing. Thus, there are delicate tradeoffs among statutory burdens of proof, required federal research budgets, the pace of innovation, and the cost and public accessibility of innovative new products.

Premarket Versus Post-Market Safety Information

In assessing whether a statute strikes a healthy balance between innovation and safety regulation, another key feature is *when* the statute requires evidence of safety to be developed. If products must be shown safe before they go on the market, this seemingly enhances safety but the costs and delays of developing premarket safety data may prevent the entry of innovative, lower-risk products. A robust program of post-marketing risk detection and analysis can relieve pressure to achieve certainty about safety prior to new product entry, thus facilitating innovation while promoting safety through rapid risk detection and response. Even when a statute requires premarket risk analysis, post-marketing risk assessment adds an important layer of protection because some risks cannot be detected in small-scale, short-duration premarket studies and can only be evaluated in the post-marketing period after products move into wide commercial use (Evans, 2009). When assessing a statute, key questions are as follows:

- Does the statute emphasize premarket safety studies, post-marketing safety surveillance and studies, or both?
- After new products enter the market, does the regulator bear the ongoing burden of detecting safety problems through inspections and testing or is there a framework for sponsors or users to report safety incidents[7] to the regulator?
- Are there mandatory requirements for product sponsors to report safety incidents after a product enters the market or is the framework voluntary?
- Does the statute equip the regulator with data infrastructure or foster the creation of data resources to support the use of observational methodologies as a tool for proactive, continuous detection of emerging risk signals (active safety surveillance) after a product enters the market?[8]
- Can the regulator require product sponsors to conduct safety studies and clinical trials in the post-marketing period to clarify signals of new safety risks?

Provisions to Foster Private-Sector Participation in Creating Evidence and Data Infrastructure for Risk Analysis and Setting Safety Standards

In regulatory frameworks, the participation of private-sector actors goes beyond simply requiring product sponsors and developers to generate risk information about their own commercial products. The work of ensuring consumer, occupational, and environmental safety at times requires broader public–private collaborations in which governmental and private actors join forces to tackle problems that are too complex for either to address alone. A host of tools already exists for orchestrating public–private collaborations that merge the activities of governmental bodies and private organizations in inventive ways (Kettl, 2002; Salamon, 2002; Yescombe, 2007). These tools include traditional instruments such as governmental grants and contracts that enable the government to engage private actors (for-profit and nonprofit) in tasks, such as risk analysis, that promote regula-

---

[7]Incident reporting allows the regulator to conduct passive surveillance by monitoring the reports it receives.

[8]An example of such data resources would be FDA's post-market risk identification and analysis system (Sentinel System), a very large-scale data resource FDA developed via a public–private collaboration for use in active post-marketing drug safety surveillance, using authorities Congress granted in the Food and Drug Administration Amendments Act of 2007 (21 U.S.C. §§ 355(k)(3), (4)).

tory objectives. However, they also include tax incentives, insurance, and loan guarantees to mobilize private capital and a wide array of other mechanisms through which governmental regulators can harness private-sector resources and know-how (Salamon, 2002) to address challenges, such as how to finance and develop large-scale, industry-wide data infrastructures that support post-marketing product safety surveillance and continuous learning or how to assemble "knowledge commons" (shared data resources) to promote innovation and rapid dissemination of best practices within an industry (Frischmann et al., 2014). In the 20th century, regulation via hierarchical governmental bureaucracy was the predominant organizational model for fulfilling public-policy goals (Goldsmith and Eggers, 2004), and this model is reflected in many of the 20th-century statutes that authorize federal agencies to seek to regulate the safety of novel biotechnology products. Alongside these statutes, however, there is a decades-long trend toward greater integration of private actors—both commercial firms and nonprofit organizations—into the day-to-day work of these agencies through public–private partnerships and other modes of collaboration (Goldsmith and Eggers, 2004).

Provisions for Managing Risks Revealed During Risk Assessment

Many statutory provisions affect risk management more than risk assessment. For example, what powers does the regulator have to respond to an emerging safety problem? Does the regulator have a nuanced set of tools that allow safety problems to be resolved while keeping beneficial products available to consumers—for example, powers to require labeling changes or to restrict use, sale, or distribution of the product as opposed to banning it? How nimble is the agency's authority to act? Are there cumbersome procedural requirements (such as having to promulgate a regulation in order to investigate or respond to a safety concern) or formal procedural requirements (such as having to conduct hearings or other formal processes before taking action)?

CONSUMER AND OCCUPATIONAL SAFETY

Diverse, new types of biotechnology products entering the market may pose new consumer- and occupational-safety challenges. The important questions for a specific product are as follows:

- Does any federal regulatory agency have jurisdiction to regulate it and, if so, which agency (or agencies)?
- Do the regulatory agencies have adequate tools to analyze the types of risk the product may present?

Multiple federal agencies are responsible for consumer and occupational safety. Under the Food, Drug, and Cosmetic Act (FDCA), FDA is a major product safety regulator. EPA is mandated to ensure safety of chemicals across a number of uses, including consumer products, occupational exposures, and manufacturing. Under the Federal Insecticide, Fungicide, and Rodenticide Act (FIFRA)[9] and Section 408 of the FDCA, EPA regulates consumer safety with respect to pesticides and pesticide residues in food and occupational safety with respect to uses of pesticides in the workplace. As shown in Table 3-1, USDA plays a crucial role in consumer safety with respect to various food products, such as meat and poultry.

Federal oversight of consumer and occupational safety is not limited to the three agencies discussed in the 2017 update to the Coordinated Framework (EOP, 2017). In addition to EPA, FDA, and USDA, the Consumer Product Safety Commission (CPSC) has residual jurisdiction to regulate consumer products not regulated by the other agencies, so biotechnology products that are

---

[9]7 U.S.C. § 136 *et seq.*

not regulated by one or more of the statutes listed in Table 3-1 may fall under CPSC's jurisdiction. Under FIFRA, EPA is completely responsible for all pesticide risks, including in the workplace. Under TSCA, EPA and OSHA share responsibility for chemical safety in the workplace; EPA evaluates and manages the chemical risk, and OSHA establishes and enforces workplace exposure limits and safety practices. In coming years, OSHA may confront novel issues in workplaces where biotechnology is used as a means of production in diverse industrial, commercial, and agricultural settings. Agencies concerned about consumer and occupational safety have to be attuned to assessing not just the types of new products but also the characteristics of the anticipated users because the safety of some future biotechnology products depends on the skill and intent of the user along with the product design (Box 3-2).

Additional agencies may need to become involved when specific biotechnology products fall within their jurisdiction—for example, the National Highway Traffic Safety Administration (NHTSA) may be called upon to address safety issues related to future biotechnology-based car batteries. This section does not attempt to identify every agency, such as NHTSA, that may occasionally encounter new biotechnology products and instead focuses on agencies—FDA, CPSC, OSHA, EPA, and finally USDA—with broad jurisdiction over consumer and occupational safety.[10]

U.S. Food and Drug Administration

FDA has various general powers to protect public health and safety, such as the authority FDA shares with the Centers for Disease Control and Prevention (CDC) under the Public Health Service Act to control the spread of communicable diseases (Hutt et al., 2014). However, FDA's authority to protect consumer safety under the FDCA is determined "almost entirely by the list of product categories over which it has jurisdiction" (Hutt et al., 2014:77); FDA protects consumers by regulating products that fall within definitions that Congress establishes. Since the 1992 update to the Coordinated Framework was issued, Congress has enacted new statutes affecting FDA's jurisdiction, including the Dietary Supplement Health and Education Act of 1994 (DSHEA)[11] affecting dietary supplements and the Family Smoking Prevention and Tobacco Control Act of 2009[12] authorizing FDA to regulate tobacco products.

For this discussion, the important FDA-regulated product categories are conventional foods (including various subcategories such as raw agricultural commodities), food ingredients that are generally recognized as safe (GRAS), food additives, medical foods, infant formula, pesticide and environmental contaminants in food, dietary supplements, cosmetics, tobacco products, new animal drugs, and animal food. Appendix D presents the statutory definitions of these terms. Whether a future biotechnology product will be regulated by FDA—and, if so, the specific risk assessment and management tools FDA can apply to that product—is determined by whether the product fits within one of these definitions. FDA makes the threshold decision whether a product fits into one of its regulated product categories, and U.S. federal courts largely—but not always[13]—defer to FDA's decisions.

Congress has granted FDA a distinct set of statutory powers to analyze and manage product safety risks for each product category. Table 3-2 summarizes the risk assessments Congress has prescribed for the various FDA-regulated categories. The discussion that follows focuses on a

---

[10]The Federal Trade Commission (FTC) plays a central role in consumer safety through its efforts to ensure that people's choices are well informed by truthful advertising and risk disclosures, but FTC is not discussed further because FTC does not directly regulate product safety or analyze product risks.

[11]Dietary Supplement Health and Education Act of 1994, P.L. 103-417.

[12]Family Smoking Prevention and Tobacco Control Act of 2009, P.L. 111-31.

[13]See, for example, *FDA v. Brown & Williamson Tobacco Corp.*, 529 U.S. 120 (2000) (rejecting FDA's assertion that cigarettes could be regulated as a medical device).

TABLE 3-2 Premarket and Post-Market Statutory Risk Assessment for Examples of Different Product Categories Regulated by the U.S. Food and Drug Administration (FDA)

Product Category	Premarket Risk Assessment[a]	Post-Market Risk Assessment
Cosmetic	Cosmetics must be safe for their intended use and properly labeled, but no premarket evidentiary review of safety by FDA is required.	FDA regulates cosmetics through post-market risk-assessment processes (e.g., inspections, analysis of samples, and enforcement of its provisions on adulteration and misbranding) and can react to safety problems, but the burden is on FDA to establish that a safety problem exists.
Food	Food does not require FDA's premarket approval, and no premarket evidence of food safety is required. FDA does not review food labels prior to marketing but has regulations that require certain types of information to be disclosed.	Congress authorizes FDA to regulate food safety mainly through post-marketing mechanisms (e.g., inspections, testing, and enforcing adulteration and misbranding standards and good manufacturing practices). FDA bears much of the burden of detecting food-safety problems, but its regulations call for food manufacturers to report certain types of safety incidents to FDA. Once a food-safety problem is detected, FDA has multiple tools to manage the problem (e.g., seizures, injunctions, criminal sanctions, warning letters, and publicity).
Food Additive	New food additives cannot be sold until FDA determines they are safe, and product sponsors (manufacturers and distributors) must produce evidence of safety. Ingredients that are generally recognized as safe (GRAS) do not require premarket approval as food additives.	Regulated as food (see above).
Dietary Supplement	Manufacturers and distributors (sponsors) must notify FDA 75 days before a new product enters the market. They must include information explaining the basis for concluding that the product is reasonably expected to be safe, but they do not have to await affirmative approval by FDA before marketing the product.	Regulated as food (see above). FDA monitors adverse event information and conducts its own research to monitor safety following approval.
Medical Foods	No premarket approval by FDA is required, but any claims in the product's labeling must be truthful and nonmisleading.	Regulated as food (see above).
Infant Formula	FDA does not approve infant formulas before they are marketed, but manufacturers must register with FDA and notify the agency before they introduce a new product, and they must comply with nutrient requirements and other regulations directed at ensuring product safety.	Regulated as food (see above).
Tobacco Product[b]	Manufacturers are subject to registration and product listing requirements. New tobacco products that make claims about modified risks are subject to FDA review and clearance or approval prior to marketing. Authority to regulate tobacco products includes related products such as e-cigarettes.[c]	FDA can set and enforce standards for adulterated and misbranded tobacco products and can restrict distribution, promotion, and advertising of tobacco products.

TABLE 3-2 Continued

Product Category	Premarket Risk Assessment[a]	Post-Market Risk Assessment
New Animal Drug	New animal drugs cannot be marketed until FDA affirmatively approves them. The burden of proof is on the manufacturer to show that the drug is safe and effective for the animal and—for drugs used in food-producing animals—that food products derived from treated animals are safe for consumption.	FDA performs inspections and requires reporting of certain safety problems by sponsors and manufacturers as well as encouraging reporting by veterinarians and animal owners of safety problems with approved animal drugs.[d] FDA has a broad array of enforcement powers to address safety problems once they are detected.
Drug (relevant insofar as it sets boundaries on other product definitions)	New drugs cannot be marketed until FDA affirmatively approves them based on evidence of safety and effectiveness. The burden of proof is on the product sponsor (the manufacturer) to provide evidence of safety and effectiveness, generated at the sponsor's expense via a three-phase premarket clinical study process that is itself regulated by FDA.	FDA has strong post-marketing regulatory powers and a broad range of tools to evaluate and manage post-marketing safety risks, and can require sponsors to report safety information and, under some circumstances, can require post-marketing studies and clinical trials. FDA has data infrastructure for both passive and active safety surveillance.
Device (relevant insofar as it sets boundaries on other product categories)	FDA uses a risk-stratified approach for clearing or approving new devices. High-risk devices must be affirmatively approved by FDA, and manufacturers face evidentiary burdens similar to those required for drugs. Moderate-risk devices must be cleared by FDA prior to marketing, but manufacturers only must show that they are substantially equivalent to a device already on the market. Many low-risk devices are exempt from premarket evidentiary review.	FDA has strong post-marketing powers to detect and manage safety risks, including general controls applicable to all devices plus additional controls aimed at higher-risk devices. Safety surveillance is primarily passive although efforts are under way to develop infrastructure to support active safety surveillance.

[a]U.S. Food and Drug Administration. Is It Really FDA Approved? Available at http://www.fda.gov/forconsumers/consumerupdates/ucm047470.htm. Accessed September 30, 2016.

[b]Hutt et al. (2014).

[c]*Sottera, Inc. v. FDA*, 627 F.3d 891 (D.C. Cir. 2010).

[d]U.S. Food and Drug Administration. How to Report Animal Drug Side Effects and Product Problems. Available at http://www.fda.gov/animalveterinary/safetyhealth/reportaproblem/ucm055305.htm. Accessed October 3, 2016.

select group of these categories that may pose special risk-assessment challenges or arouse high public concern and concludes with a review of FDA's role in product-development research. FDA's regulation of human drugs and medical devices lies outside the scope of this report. Nevertheless, drugs and devices are relevant in two situations: (1) when FDA's discretionary decision to treat a product as a drug or device removes the product from regulation as a food, food additive, cosmetic, or other FDA-regulated category and (2) when FDA's discretionary decision to categorize a product as a drug or a device removes the product from CPSC's jurisdiction.

Biotechnology-Based Cosmetics

Cosmetics are "the least intensively regulated of all the products under FDA's jurisdiction" (Hutt et al., 2014:110). Congressional legislation was proposed in 2016 to strengthen FDA's framework for cosmetics, but no amendments had been adopted or implemented at the time the committee was writing its report. Thus, this discussion explores whether FDA could effectively regulate novel biotechnology cosmetics with the powers it had at the start of 2017.

Because its existing framework for cosmetics risk assessment is weak, FDA has regularly used its discretion to deem cosmetics to be drugs or devices. The mere fact that a cosmetic poses a safety risk does not itself place the product within FDA's drug or device definitions. FDA can, however, regulate a cosmetic as a drug or device if the product sponsor displays intent for the cosmetic to be used for the "cure, mitigation, treatment, or prevention of disease" or if the cosmetic is "intended to affect the structure or any function of the body" (see Appendix D). These concepts are broad enough to encompass many future biotechnology-based cosmetics.

Deeming a cosmetic to be a drug lets FDA require premarket proof of safety and effectiveness and subjects the product to all of FDA's strong pre- and post-marketing drug regulatory powers. A problem with this approach is that FDA's drug regulation is not risk stratified; that is, it requires product sponsors to bear the costs and delays of going through FDA's premarket clinical trial process for all new drugs, regardless of the level of risk they pose. This lack of differentiation could deter the development of innovative cosmetic products. There is no middle ground between FDA's cosmetic regulation (which provides no premarket regulatory review) and FDA's drug regulation (which subjects all drugs to the same premarket review process FDA requires for high-risk cancer therapies).

Some biotechnology-based cosmetics may qualify as medical devices rather than drugs. A product can be categorized as a device, rather than a drug, if the product "does not achieve its primary intended purposes through chemical action within or on the body . . . and is not dependent upon being metabolized" (see Appendix D). When applicable, FDA's device regulations offer a nuanced, risk-stratified review process that could ensure consumer safety with fewer impacts on beneficial innovation.

FDA noted in a 2011 draft guidance that many products lend themselves to characterization either as a drug or a device (FDA, 2011), which will often be the case with future biotechnology products, especially those that achieve their effects at a genomic or microscopic scale where the mode of action could legitimately be characterized as either chemical/metabolic (drug) or mechanical/electrical (device) (Evans, 2015a). This guidance provides FDA with significant discretion to categorize particular products variously as cosmetics, drugs, or devices. Through careful exercise of this discretion, the agency has significant power to position products under the risk-assessment framework best suited to the task of protecting human safety while still fostering beneficial innovation.

Biotechnology Foods, Food Additives, and Dietary Supplements

Food products derived from genetically engineered (GE) or genome-engineered plants and animals ("biotechnology foods") were already foreseen in the 1980s, and FDA has policies in place to address such products, which were discussed in the 2017 update to the Coordinated Framework (EOP, 2017). Future biotechnology products may include an additional array of new products, for example, synthesized foodstuffs produced directly in industrial and fermentation facilities without the intermediation of plants or animals (such as egg-white protein produced from GE yeast) or cultured food products like yogurt containing GE microorganisms. Some of these future foodstuffs may fit within FDA's existing policies, but others may present challenges. This discussion examines the flexibility of FDA's statutes to cope with such products.

Consumer Information. One matter of public concern is whether consumers will receive information to guide their decisions about exposing themselves to biotechnology food. FDA shares responsibility with the other Coordinated Framework agencies to ensure safety of the human food supply. Food is circularly defined in the FDCA as "articles used for food or drink" and components thereof (see Appendix D). Courts often follow a commonsense approach when assessing whether a product is a food.[14] In the early and mid-20th century, FDA aggressively required manufacturers of imitation foods to label their foods as such to prevent consumer deception. Since the 1970s, however, FDA has required substitute or synthetic foods to be labeled as imitations only if they are nutritionally inferior to the food they resemble (Hutt et al., 2014).[15] Thus, synthetic biotechnology foods would not be required to disclose "imitation" status if nutritional equivalency is established.

With respect to foods derived from GE plants, FDA does not consider the mere fact of a modification to be a "material fact" that must be disclosed in food labeling,[16] even if consumers have a strong desire to know. FDA requires disclosure only if there is a food quality or safety issue, and FDA bears the burden of substantiating such issues (FDA, 1992).[17] Thus, FDA's food labeling statutes do not ensure consumers will be informed when they are exposing themselves to biotechnology food products.

In July 2016, President Obama signed a bill amending the Agricultural Marketing Act of 1946 to require USDA to establish labeling requirements for food products containing bioengineered or genetically modified organisms (Fama, 2016). Labeling would apply to many, but not necessarily all, types of future biotechnology food. Legal commentators have noted that the law may not cover some bioengineered foods, for example, foods that are the product of genetic deletions (Elliott, 2016). USDA was given 2 years to promulgate regulations and determine the thresholds for biotechnology-derived ingredients in food that would trigger a disclosure requirement. Once promulgated, these federal regulations will preempt the biotechnology food labeling requirements that some states have enacted.

Premarket Safety Review. A second matter of public concern is whether biotechnology-derived foods will receive premarket safety review. FDA regulates food safety mainly through post-marketing mechanisms, such as inspections, testing, issuing good manufacturing practice regulations, and enforcing FDA's prohibitions on commerce in adulterated and misbranded foods. Once a safety problem is detected with a marketed food product, the agency has strong enforcement tools such as product seizure, injunctions, civil and criminal sanctions, or issuing warnings and publicity. However, FDA bears much of the burden to detect food-safety problems, although the FDCA imposes some duties for food manufacturers and suppliers to report safety issues under some circumstances.

As a general matter, FDA does not subject food to any premarket safety review or require manufacturers to submit evidence of safety before new products can be sold. The FDCA does, however, define several subcategories of food—medical foods, infant formula, food additives, and dietary supplements—that are subject to special regulatory requirements. For food additives and dietary supplements, at least some premarket evidence of safety is required:

- **Food additives.** New food additives cannot be sold until FDA determines they are safe, and product sponsors must produce evidence of safety and await FDA approval before they can be marketed. Ingredients that are GRAS do not require premarket approval as food additives. The FDCA provides a strong evidence-forcing mechanism that places the burden

---

[14]See, for example, *Nutrilab, Inc. v. Schweiker,* 713 F.3d 335 (7th Cir. 1983) (discussing the statutory definition of food).

[15]See also Federal Food, Drug, and Cosmetic Act § 403(c), 21 U.S.C. § 343(c) (providing that "a food shall be deemed to be misbranded if it is an imitation of another food unless its label bears, in type of uniform size and prominence, the word 'imitation' and, immediately thereafter, the name of the food imitated") but see 21 C.F.R. § 101.3(e) (focusing FDA's inquiry on whether a substitute food is nutritionally inferior).

[16]21 U.S.C. § 321(n).

[17]See also *Alliance for Bio-Integrity v. Shalala*, 116 F.Supp.2d 166 (D.D.C. 2000).

of proof on food additive manufacturers to prove their food additives are safe. FDA's 1992 Statement of Policy on Foods Derived from New Plant Varieties (FDA, 1992) deemed foods from GE plants to be GRAS, thus ceding the agency's strong statutory evidence-forcing mechanism with regard to biotechnology foods, including food additives produced through biotechnology. FDA reasoned that the addition of genetic material (nucleic acids) to foods is GRAS because nucleic acids already exist in all plant and animal foods used by humans. Concerns could arise only if the added genetic material expresses a protein or substance that differs significantly from substances already found in the food. Thus, this GRAS presumption is rebuttable, and FDA can still require a sponsor to submit a food additive petition and require premarket approval if the genetic manipulation transfers genes from a species that is a known food allergen or causes the food product to contain a novel protein that arouses safety concerns. Without an evidence-forcing mechanism, however, the burden is largely on FDA to develop evidence with which to rebut its GRAS presumption. FDA has implemented a process for voluntary consultation prior to market entry for such products (FDA, 1997) and reports wide industry adherence to the process (EOP, 2017).

- **Dietary supplements.** Manufacturers wishing to market "new" dietary ingredients (that is, ingredients not already marketed in the United States before 1994 and which have not previously been in the food supply as articles used for food without chemical alteration) must give notice to FDA at least 75 days before introducing the product. This notice must include information that supports the conclusion that a supplement containing the ingredient can reasonably be expected to be safe (Levitt, 2001). This evidence-forcing mechanism requires sponsors to provide some evidence of safety, but they do not have to await affirmative approval by FDA before marketing the product once the 75-day notice period has expired, which "makes it essential for public health protection that FDA have the resources to review the notifications in a timely manner" (Levitt, 2001). Once on the market, a dietary supplement is subject to FDA's broad prohibition on the sale of adulterated food. A supplement is considered adulterated if it (or one of its ingredients) presents "a significant or unreasonable risk of illness or injury" when used in accordance with its label, or under normal conditions of use if there are no directions in its label. "The burden of proof is on FDA to show that a product or ingredient presents such a risk" (Levitt, 2001).

Even though there is no general requirement for FDA premarket safety review of food, the statutes just described allow at least some premarket safety review of food additives and dietary supplements. FDA thus has various tools, under its existing statutes, to address safety concerns that novel biotechnology food products may present in the coming 5–10 years. First, if FDA determines that it needs to adapt its 1992 policy and consultation process for food derived from GE or genome-engineered plant varieties, the agency can do so without any further congressional action. Second, the 1992 policy, which deemed many biotechnology foods to be GRAS, only applies to foods derived from GE plant varieties. Some of the new biotechnology foods coming to market may lie outside that policy and, thus, outside the 1992 GRAS presumption. For example, a synthesized spice or flavoring ingredient or a synthetic egg used as an ingredient in processed foods may fit within FDA's general definition of a food additive (see Appendix D), requiring a food additive petition and premarket safety review. The 1994 DSHEA defines dietary supplements as including, in addition to vitamins and minerals, "a dietary substance for use by man to supplement the diet by increasing the total dietary intake" (see Appendix D, 21 U.S.C. § 321(ff)(1)(e)). This broadly worded clause may allow FDA to treat some novel biotechnology foods as dietary supplements, requiring sponsors to notify FDA of their plans to market the product and explain their basis for concluding that the product can reasonably be expected to be safe. FDA thus has various existing tools for fashioning case-by-case solutions that protect consumer safety without unduly burdening innovation.

Validity of Nutritional and Health Claims. FDA regards safety as a balance of benefits and risks. If false or unsubstantiated claims about nutritional and health benefits are made about biotechnology food products, this potentially poses a consumer-safety risk. An important part of risk assessment, therefore, is to ensure that any nutritional and health claims about biotechnology foods are accurate—or, if they are scientifically uncertain, to ensure the uncertainty is properly disclosed. FDA has several ways to protect consumers from false or misleading claims about foods and dietary supplements, including those that are derived from or produced through biotechnology.[18] First, the 1990 Nutrition Labeling and Education Act (NLEA) authorized FDA to review evidence to support health claims about foods and dietary supplements. Product sponsors can submit a health claim petition to FDA and supply evidence to support the claim, or FDA can initiate a review on its own. Second, the 1997 Food and Drug Administration Modernization Act allows health claims that are supported by an authoritative statement by the National Academy of Sciences or a scientific body of the U.S. government with responsibility for public health protection or nutrition research. Food and supplement manufacturers can make such claims after a 120-day notification to FDA, unless FDA indicates the notification is deficient. There is a third pathway for claims that do not meet FDA's "significant scientific agreement" standard for validity of health claims. The First Amendment prevents FDA from blocking such claims entirely, but FDA can require disclosure that the claims are uncertain. FDA has published guidance about appropriate disclosures (FDA, 2003). Finally, FDA has authority to deem a food or supplement to be a drug if the product makes therapeutic claims that display the sponsor's intent to market the product as a drug (see Appendix D). By enacting NLEA and DSHEA, Congress struck a balance between safety and innovation that constrains FDA's authority to subject foods and dietary supplements to the full rigor of FDA's premarket drug-approval process.[19] Still, the agency is not without power to deem a novel biotechnology product to be a drug if the claims made about it and other circumstances make that course appropriate. Thus, FDA has multiple risk-assessment tools available to protect consumers from false or misleading claims about the health benefits of new biotechnology foods.

Summary of FDA's Authorities for Food-Safety Risk Assessment. The trends identified in Chapter 2 present challenges and counsel a need to ensure that FDA's Center for Food Safety and Applied Nutrition and Center for Veterinary Medicine receive adequate resources for the task ahead. Some of these trends, such as accelerated product development and increased scale of new product entry, may affect the volume of workload. The premarket notification framework of the dietary supplement statute, as noted above, only protects consumers if FDA is adequately staffed to review and respond to notifications in a timely way.

An additional concern is the growth of do-it-yourself biology (DIYbio), small-scale, and decentralized product development. FDA's safety oversight framework relies, in substantial part, on having compliance-oriented regulated companies that can meet the agency halfway in ensuring consumer safety. In future years, FDA may face the challenge of regulating small product developers that lack internal regulatory compliance resources and require an added level of consultation and education by the agency. For example, will FDA's voluntary consultation process for GE crops and food derived from GE crops, which has attracted a high level of participation by traditional manufacturers, elicit similar rates of participation by individual DIYbio and small manufacturers? Will

---

[18]U.S. Food and Drug Administration. Label Claims for Conventional Foods and Dietary Supplements. Available at http://www.fda.gov/food/ingredientspackaginglabeling/labelingnutrition/ucm111447.htm. Accessed September 30, 2016.

[19]As noted by Hutt et al. (2014:103), "the most important aspect of DSHEA is that it permits a dietary supplement to make a structure/function claim without rendering itself a food, even if the supplement is not a 'common sense' food."

FDA's traditional inspection processes be strained as a growing decentralization of food-product development allows individual growers to harness DIYbio to modify their own small crops? Additional outreach and leveraging of FDA's own regulatory resources—including but not necessarily limited to measures already envisioned in the 2017 update to the Coordinated Framework—will be important to meet such challenges.

Biotechnology Animals

FDA currently regulates most GE animals under the FDCA's new animal drug provisions by treating genetic material that is integrated into the animal as a new animal drug (FDA, 2015a, 2017a).[20] FDA's new animal drug risk assessment considers a drug's safety and effectiveness to the animal and, in the case of food-producing animals, whether food derived from the animal is safe for consumption. The 2017 update to the Coordinated Framework describes FDA's programs for protecting consumers from risks from eating food derived from GE animals (EOP, 2017).

Biotechnology-altered animals of the future may include nonfood animals, such as pets or species brought back from extinction. In addition to environmental risks, these animals may pose consumer-safety risks. For example, a biotechnology-altered pet could have altered susceptibility to zoonotic diseases or aggressive traits that pose injury risks to humans. According to the FDCA, the term *safe*, as used in the new animal drug provisions, "has reference to the health of man or animal."[21] While this is sometimes conceived as an authority merely to ensure the safety of foods derived from food-producing animals, it actually carries a broader authority to consider human-safety impacts of new animal drugs. Thus, if an alteration to an animal results in human risks that go beyond food-safety risks, FDA has authority to take these other risks into account as part of a user-safety evaluation.

The specific risks are, however, difficult to foresee in advance and a critical dimension of consumer protection is to have robust post-marketing systems for prompt detection of emerging problems and rapid response if they arise. FDA and CDC share authorities under the Public Health Service Act, unrelated to FDA's regulation of new animal drugs, that empower them to respond to outbreaks of communicable diseases caused by familiar and exotic species. An example was their coordinated response to an outbreak of monkeypox caused by traffic in exotic species in 2003 (Crawford, 2003). The first line of defense is CDC's regular public health surveillance activities, which monitor communicable disease outbreaks, including zoonotic diseases, and which also monitor injuries to humans caused by animals (CDC, 2003). FDA and CDC then have powers to coordinate a response to protect the public. Upon detecting the monkeypox outbreak, FDA and CDC issued a joint order to other federal agencies, including the U.S. Department of Transportation, and to state agriculture and health agencies including state and public health veterinarians and state fish and wildlife officials, among others (Crawford, 2003).

A more speculative concern is the possibility that biotechnology might be used to engineer deleterious traits into animals or that deleterious traits might be introduced as a side effect of genetic manipulation. CDC, joined by the American Society for the Prevention of Cruelty to Animals,[22] has taken the position that species- or breed-specific regulations (for example, regulation of pit bulls) are an ineffective way to address problems such as human injuries caused by animals. They suggest that a more effective approach is to address underlying societal concerns, such as the existence of industries (for example, dog fighting) that promote the inappropriate enhancement of deleterious

---

[20]In addition to FDA's authority to regulate new animal drugs under the Federal Food, Drug, and Cosmetic Act, USDA has authority to regulate biologic medicines used in animals under the Virus-Serum-Toxin Act of 1913 and related statutes.

[21]21 U.S.C. § 321(u).

[22]See American Society for the Prevention of Cruelty to Animals, Breed Specific Legislation. Available at http://www.aspca.org/animal-cruelty/dog-fighting/what-breed-specific-legislation. Accessed September 30, 2016.

traits in animals. For animals that are the products of biotechnology, a similar approach may be appropriate: protecting the public will require an appropriate dedication of resources for CDC to conduct surveillance for rapid detection of animal-related injuries as well as zoonotic disease outbreaks traceable to new varieties of biotechnology-altered animals. If signals of a problem are detected, FDA, CDC, and other state and federal agencies have powers to address them through existing legal approaches. Underlying social problems that may cause people to engineer animals for inappropriate goals will need to be tackled directly using means such as publicity, education, and public engagement to forge consensus around norms that protect animals as well as the people exposed to them.

Product-Development Research

FDA regulates new drugs and devices as an Investigational New Animal Drug (INAD),[23] Investigational New Drug (IND),[24] or Investigational Device Exemption (IDE)[25] to allow the agency to restrict the distribution and use of unapproved animal drugs, human drugs, and medical devices. Human drug and device regulation is outside the scope of this report, yet these regulations deserve mention for two reasons. First, as already discussed, the definitional lines are such that some cosmetics, foods, and other consumer products may, at times, fall within FDA's drug or device definitions. When this is so, product-development research may lie within the reach of the IND and IDE regulations. Second, the INAD, IND, and IDE regulations may have growing importance in the face of a trend, described in Chapter 2, toward diversification of the financing sources for biotechnology-product development. Diffusion of research to DIYbio and smaller, noninstitutional research settings potentially may move research outside the centralized, NIH-led research oversight process that has been a central pillar of the Coordinated Framework.

The NIH guidelines detail safety practices and containment procedures for basic and clinical research involving recombinant or synthetic nucleic acid molecules, including the creation and use of organisms and viruses containing recombinant or synthetic nucleic acid molecules. The NIH guidelines apply to research funded by NIH and various other federal agencies but are voluntarily followed by some other research organizations. Even when the 1992 update to the Coordinated Framework was first published, contemporary commentators expressed concern about the possibility that "unregulated organisms" might fall outside USDA and EPA's authorities and might also escape the NIH guidelines if developed at a research institution or with funding not subject to those guidelines. The evolving structure and financing of biotechnology research refresh this concern.

FDA's INAD, IND, and IDE regulations offer potentially important pathways to ensure appropriate federal regulatory oversight of some categories of research that may fall outside the NIH guidelines. For example, FDA's IND and IDE regulations define "sponsor-investigators" as individuals who both initiate and conduct (alone or with others) an investigation and under whose direction the experimental product is administered, dispensed, or used.[26] Many at-home or DIYbio enthusiasts fit within this concept. Sponsor-investigator studies are thought to present special risks to humans exposed to non-FDA-approved products because merging the sponsor and investigator roles removes a layer of checks and balances that ordinarily exist in research (TPG, 2005). Because of this concern, FDA's training materials suggest that an FDA-approved IDE or IND may be required for *any* sponsor-investigator study of an unapproved product, "even if no marketing application is planned" (Henley, 2013)—that is, even if the sponsor-investigator has no plans to commercialize the product for wider sale. The structure and function clauses of FDA's drug and

---

[23]21 C.F.R. § 511.1.

[24]21 C.F.R. § 312.

[25]21 C.F.R. § 812.

[26]21 C.F.R. § 812.3(o) [devices]; 21 C.F.R. § 312.3 [drugs].

device definitions, as already discussed, grant the agency significant authority to deem products to be investigational drugs or devices if they are intended to affect the structure or function of the body (see Appendix D). Requiring at-home and DIYbio enthusiasts to obtain an IND or IDE for certain categories of research is one possible mechanism to ensure that the protocols and human-subject protections receive external oversight by FDA (Evans, 2015b). Statutory authority thus may exist for FDA to oversee some categories of at-home and DIYbio experiments that otherwise threaten to escape federal research oversight. This role would, however, imply a significant expansion of workload for FDA and would require an appropriate expansion of staffing and resources.

Consumer Product Safety Commission and the U.S. Food and Drug Administration

The growing list of product categories potentially transformed by biotechnology requires a discussion of how FDA's consumer product regulation interacts with CPSC's regulatory role. The CPSC has jurisdiction over "consumer products," which broadly encompass articles sold and distributed to consumers.[27] The statutory definition of CPSC-regulated products expressly excludes products regulated by FDA, EPA, and USDA. CPSC's jurisdiction to regulate biotechnology consumer products thus is a residual, gap-filling jurisdiction over products that fail to fall under the jurisdiction of other Coordinated Framework agencies. CPSC does not regulate pesticides,[28] tobacco products,[29] drugs, devices, or cosmetics subject to FDA regulation,[30] or food.[31] As a result, FDA's decisions about whether products fit into its own jurisdiction can have the effect of shrinking or expanding the list of products that CPSC regulates.

A frequently discussed example is FDA's power to deem a product to be a medical device if it is "intended to affect the structure or any function of the body of man or other animals" (Appendix D, 21 U.S.C. § 201(h)(3)). Construed broadly, this definition would allow FDA to assert jurisdiction over a wide range of consumer products like air conditioners, shoes, and sporting goods that arguably affect the structure and function of the human body. FDA has traditionally construed the definition narrowly, limiting itself to products that have a medical or therapeutic impact and leaving other products to CPSC. Nevertheless, a former FDA chief counsel stated that "if Section 201(h)(3) of the [FDCA] were interpreted to give FDA jurisdiction over any product foreseeably having an effect on the structure or a function of the body, then regulatory authority would shift from the CPSC to FDA for a host of non-health-related products."[32]

This fact may provide a useful safety valve, working within existing statutes, to protect consumer safety as the array of new biotechnology products expands to encompass household goods, computing products, clothing, cosmetics and personal care products, recreational products and

---

[27]See 15 U.S.C. § 2052(a)(5), providing, "The term 'consumer product' means any article, or component part thereof, produced or distributed (i) for sale to a consumer for use in or around a permanent or temporary household or residence, a school, in recreation, or otherwise, or (ii) for the personal use, consumption or enjoyment of a consumer in or around a permanent or temporary household or residence, a school, in recreation, or otherwise."

[28]15 U.S.C. § 2052(a)(5)(D), excluding pesticides as defined by the Federal Insecticide, Fungicide, and Rodenticide Act, 7 U.S.C. § 136 *et seq.*

[29]15 U.S.C. § 2052(a)(5)(B). See P.L. 111-31 (Family Smoking Prevention and Tobacco Control Act of 2009) placing tobacco products under FDA's jurisdiction.

[30]15 U.S.C. § 2052(a)(5)(H), excluding drugs, devices, and cosmetics as defined in section 201(g), (h), and (i) of the Federal Food, Drug, and Cosmetic Act, 21 U.S.C. § 321(g), (h), and (i).

[31]15 U.S.C. § 2052(a)(5)(I), excluding food as defined in section 201(f) of the Federal Food, Drug, and Cosmetic Act, 21 U.S.C. § 321(f); excluding poultry and poultry products as defined in section 4(e) and (f) of the Poultry Products Inspection Act, 21 U.S.C. § 453(e) and (f); excluding meat, meat food products as defined in section 1(j) of the Federal Meat Inspection Act, 21 U.S.C. § 601(j); and excluding eggs and egg products as defined in section 4 of the Egg Products Inspection Act, 21 U.S.C. § 1033.

[32]See letter from Daniel E. Troy, FDA Chief Counsel, to Jeffrey N. Gibbs, concerning Applied Digital Solutions VeriChip products (Oct. 17, 2002), reprinted in Hutt et al. (2014:125–128).

toys, and pet-care products. Scholars question whether CPSC has an adequate set of statutory tools under the Consumer Product Safety Act to cope with future biotechnology consumer products (see, for example, Lin, 2013). CPSC's authorities are weak, with heavy reliance on information disclosure and standard setting as tools of consumer protection. The agency first must rely on voluntary standards but can impose mandatory safety standards when "reasonably necessary" to address unreasonable consumer safety risks,[33] and CPSC ultimately can ban a dangerous product (Lin, 2013). The "reasonably necessary" threshold for taking mandatory action places CPSC under a burden to develop evidence of a safety problem before it can act—a burden "at least as difficult to meet as the standards imposed by the [pre-2016] TSCA" (Lin, 2013:92). CPSC has no authority to require premarket safety testing and generally finds itself in a reactive posture of responding to reports of product-related injuries after products already are in wide use. Commentators also express concern about CPSC's resources and expertise to regulate products incorporating new and emerging technologies (Felcher, 2008).

A broad reading of FDA's product definitions offers a potential pathway—under existing statutes—to bring high-risk biotechnology consumer products under FDA's jurisdiction on a case-by-case basis. FDA could continue to interpret its definitions conservatively, as it has been doing, with respect to traditional products, but might justify a broader reading for novel biotechnology consumer products where CPSC's risk-assessment tools are inadequate to protect the public. FDA does not, of course, have unlimited discretion to deem a product to lie within FDA's jurisdiction; the product must reasonably fit into one of the available FDA product categories (see Appendix D). However, as Hutt et al. (2014:125) noted, the full breadth of the structure and function clause of FDA's device definition "remains an open issue." FDA thus may have unutilized authority to expand its reach to protect the public from novel biotechnology consumer products that call for more risk assessment than CPSC is able to provide.

Occupational Safety and Health Administration

Many future biotechnology products are intended not as final consumer products but instead would be used as means of production in diverse industrial, commercial, and agricultural settings. Such products may pose risks primarily to workers rather than to consumers of the final marketed products, in which case OSHA would be the primary safety regulator. As previously discussed, EPA shares jurisdiction with OSHA to regulate occupational safety, and OSHA was part of the 1986 Coordinated Framework. When the Coordinated Framework was originally developed in 1986, OSHA concluded that it had an adequate and enforceable basis to protect the safety and health of biotechnology workers under the general duty clause, which requires employers to provide a place of employment that is "free from recognized hazards that are causing or likely to cause death or serious physical harm."[34]

Several concerns surround OSHA's ability to regulate the safety of industrial biotechnology applications (Lin, 2013). OSHA has authority to set and enforce standards for workplace safety by setting permissible exposure limits (PELs) for hazardous materials and by establishing measures (for example, requiring protective equipment or engineering controls) to help comply with PELs.[35] However, key court decisions, by placing the burden of proof on OSHA, including the quantification of risk, have made it difficult for OSHA to set standards in situations where there is uncertainty

---

[33]15 U.S.C. § 2057.

[34]See Section 5(a)(1) of the Occupational Safety and Health Act (29 U.S.C. § 651 *et seq.*) (29 U.S.C. § 654(a)(I), see also §§ 651–678).

[35]29 U.S.C. § 655(b)(5), (7).

or limited information about the potential risks, as will often be the case when regulating future industrial biotechnologies.[36]

Another concern is whether the processes required by the Occupational Safety and Health Act may be too cumbersome to support nimble response to emerging risks. For example, with regard to nanomaterials in the workplace, some commenters have questioned whether the formal process of establishing exposure limits "could overwhelm" the capabilities of the National Institute for Occupational Safety and Health and OSHA (Bartis and Landtree, 2006). Because of these various judicial, procedural, and political constraints, some commenters feel that regulation under the Occupational Safety and Health Act will have difficulty protecting workers effectively from risks related to novel biotechnology products (Mendeloff, 1988; Shapiro and McGarity, 1989; Lin, 2013).

In summary, the Supreme Court interprets the Occupational Safety and Health Act as placing the burden of proof on OSHA, and OSHA's process of setting exposure limits is, by law, procedurally cumbersome. Existing statutes thus may make it difficult for OSHA to respond nimbly to new risks arising in the biotechnology-enabled workplace of tomorrow, and this concern is exacerbated when there is uncertainty or incomplete information about emerging risks.

U.S. Environmental Protection Agency

As discussed above, EPA's authority as it relates to consumer and occupational safety is granted under FIFRA, FDCA, and TSCA. EPA regulates consumer safety with respect to pesticides and pesticide residues in food[37] and occupational safety with respect to uses of pesticides in the workplace under FIFRA. TSCA straddles the boundaries among consumer-safety, occupational-safety, and environmental regulation, promoting all three objectives. A key point is that the regulation of "new chemical substances" described by TSCA before the commercialization of biotechnology products now includes biotechnology products, such as GE and genome-engineered organisms (EPA, 1997). TSCA authorizes EPA to regulate a diverse array of consumer, commercial, and industrial biotechnology products, to the extent these are not otherwise regulated by FDA or by EPA under FIFRA. EPA regulates the importation, production, distribution, use, and disposal of "new chemical substances," with the goal of protecting human health and the environment from unreasonable risk of injury. As enacted in 1976, the "old" TSCA had defects that caused it to be widely viewed as a dysfunctional risk-assessment framework (Culleen, 2016; Rothenberg et al., 2016). It provided for premarket notification of new chemical substances and certain significant new uses of chemicals. Products did not have to await affirmative approval and could enter the market unless EPA could bear the evidentiary burden of finding significant risk. EPA could require manufacturers to test chemicals for health and environmental effects, but EPA was hamstrung by a requirement to use notice-and-comment rulemaking (a cumbersome and slow procedure) to require such testing (Duvall et al., 2016). Moreover, the 1976 TSCA circularly required EPA to develop evidence that a chemical posed a risk before the agency could require manufacturers to conduct testing to discover what the risks were.[38] EPA could require periodic reporting of certain information, but TSCA protected the confidentiality of trade secrets in ways that blocked information flows that could have informed the public and enhanced public safety (Culleen, 2016). The 1976 TSCA

---

[36]*Industrial Union Department, AFL-CIO v. American Petroleum Institute* (448 U.S. 607 (1980)) placed the burden of proof on OSHA to show that regulation is "reasonably necessary and appropriate to remedy a significant risk of material health impairment" before it can promulgate a new standard. That case held that OSHA needed to quantify the risk of and show that it surpassed a numerical threshold of significance before OSHA could justify regulating a known carcinogen, benzene.

[37]Federal Food, Drug, and Cosmetic Act § 408, codified at 21 U.S.C. § 346a.

[38]Environmental Defense Fund. A new chemical safety law: The Lautenberg Act. June 22, 2016. Available at https://www.edf.org/health/new-chemical-safety-law-lautenberg-act. Accessed January 31, 2017.

had provisions to preempt state regulation, but they were seldom used because EPA did not enact many regulations under the original TSCA. The act's principal risk management provision (Section 6) was so unworkable that EPA proposed no rulemakings under it after a court invalidated EPA's attempt to regulate asbestos in 1991 (Duvall et al., 2016). As a result, some states implemented more stringent protections, resulting in nonuniform requirements that have been burdensome for business and innovation (Rothenberg et al., 2016).

The 2016 TSCA amendments seek to improve the risk-assessment framework. They expand EPA's authority under Section 5 of TSCA to order testing to review premarket notifications or notices of significant new uses—providing a more workable evidence-forcing mechanism than the past requirement to use rulemaking. They require EPA to make an affirmative determination that new chemicals do not present unreasonable risk of injury before they can enter commerce. The amendments also strengthen EPA's post-market authorities. They call on EPA to conduct a risk evaluation of all chemicals already existing in commerce and allow the agency to do so without first having to make legal findings that, in the past, impeded such review. The amendments tasked EPA to quickly (within a few years) promulgate regulations to govern the risk-evaluation process and to identify high-priority substances (10 within 6 months and 20 within 3.5 years) for evaluation. Once initiated, evaluations must be completed in 3 years. The statutory standard for evaluation has been modified to focus on risks of injury to humans or the environment, without having to balance costs or benefits. At the point of deciding how to manage risks, however, costs and benefits can be considered, and the earlier requirement to adopt the "least-burdensome" regulatory restrictions has been eliminated. The amendments allow EPA to collect higher fees to defray up to 25 percent of the costs of the new programs. They also promote greater transparency by allowing EPA to share confidential business information with state and tribal governments, health and environmental professionals, and first responders, although the security procedures required of recipients may still limit transparency. They implement a partial preemption scheme that grandfathers some past state laws and regulatory actions—thus leaving some ongoing lack of uniformity but moving to a higher level of uniformity prospectively.

The effects of the 2016 TSCA amendments will be felt during the 5–10-year time frame addressed in this report, but it would be incorrect to assume a sudden transformation. The roster of companies affected by TSCA will expand from traditional chemical manufacturers and processors to "potentially any manufacturer that incorporates chemicals in its products," such as manufacturers of personal care products, automotive components, computer and electronics, toys, and clothing (Sidley, 2016). Thus, developers of biotechnology-derived chemicals will feel the effects of the amended TSCA, but the question is how soon. Legal commentators caution that "it may take years for EPA to fully implement the amended law's numerous new requirements" (Culleen, 2016). At the time the committee was writing its report, key parameters still had to be worked out in series of rulemakings that were likely to span several years. Attorneys suggested that "the industry should carefully monitor new EPA regulations and decisions that may warrant judicial review" (Sidley, 2016), raising the prospect of legal challenges that could delay implementation further.

Commentators also expressed concern that TSCA's new fees and enhanced premarket review requirements could adversely affect innovation by slowing the time to market for new products, "given how few people EPA has working in the new chemicals program, and how overwhelmed they are already" (Culleen, 2016). Appropriate staffing and resources are thus crucial, both to reap the full benefits that the 2016 amendments offer and to avoid regulatory bottlenecks that threaten to slow innovation as the flood of new products anticipated in Chapter 2 confronts TSCA's new premarket-notification requirements.

U.S. Department of Agriculture

USDA's authority for consumer safety[39] rests within the Federal Meat Inspection Act,[40] the Poultry Products Inspection Act,[41] and the Egg Products Inspection Act.[42] Pursuant to these statutes, USDA's Food Safety and Inspection Service (FSIS) inspects meat, poultry, processed eggs, and certain fish moving in interstate commerce. USDA and FDA have a long history of coordinating their complex shared jurisdiction to regulate food safety. Thus, for example, FDA has exclusive regulatory jurisdiction over live animals intended to be used for food, but USDA oversees slaughter and processing of meat and poultry, and USDA has ceded jurisdiction to FDA for food products containing less than 2 percent of meat and poultry content (Hutt et al., 2014). The agencies also coordinate with respect to food products that fall within their shared jurisdiction, which would continue under the 2017 update to the Coordinated Framework (EOP, 2017). According to that update, FDA will apprise FSIS of reviews concerning the safety of meat, poultry, eggs, and fish of the Order of Siluriformes from GE animals if there is intent to use the animals for food production. FDA also will advise FSIS about FDA's assessments concerning the safety of substances added to food animals via genetic engineering. FSIS will evaluate whether the addition of such substances is permissible under its own statutes and will communicate with the public and other stakeholders. Synthetic food products, such as synthesized meat and eggs produced without the intermediation of animals, appear to fall outside the definitions of non-GE meat and egg products that FSIS regulates and responsibility for the safety of such products would lie primarily with FDA.

ENVIRONMENTAL PROTECTION

This section discusses environmental regulation of biotechnology products by EPA under FIFRA and TSCA and by USDA under the Plant Protection Act and Animal Health Protection Act (Table 3-3).[43]

U.S. Environmental Protection Agency

EPA regulates the sale, distribution, and use of pesticides under FIFRA, and these responsibilities include products produced through biotechnology. Thus, a plant-incorporated protectant (for example, a *Bt* toxin) is subject to EPA's pesticide regulations. EPA regulates the market entry of new pesticides by requiring registration of new products. Registration requires an evidence-based premarket review in which product sponsors submit evidence to demonstrate that the product will not cause unreasonable adverse effects on the environment under its proposed conditions of use. EPA facilitates the development of information to support product registration by allowing unregistered products to be field tested under Experimental Use Permits (EUPs). EUPs allow limited distribution and use of the product, under EPA oversight, for the purpose of generating data to support product registration. EPA can register products with unrestricted or restricted use or marketing based on the determined potential risk, and it has the authority to address concerns that arise after a product is already on the market.

Under TSCA, EPA has the authority to regulate the commercial use of intergeneric microorganisms, which are formed from organisms in different genera or with synthetic DNA not from

---

[39]USDA does not have authority to address occupational safety.

[40]21 U.S.C. § 601 *et seq.*

[41]21 U.S.C. § 451 *et seq.*

[42]21 U.S.C. § 1031 *et seq.*

[43]FDA is not directly involved in environmental regulation. Thus, for example, FDA only regulates biopharming—the modification of plants to express pharmaceutical products—to ensure that the resulting *drug* products are safe and effective. Regulating the uncontained testing and use of the modified plant itself is the responsibility of USDA.

TABLE 3-3 Environmental Protection Responsibilities of the U.S. Environmental Protection Agency (EPA) and the U.S. Department of Agriculture (USDA) for Biotechnology Products

Agency	Product	Statute	Authority
EPA	Pesticide	Federal Insecticide, Fungicide, and Rodenticide Act (FIFRA) Federal Food, Drug, and Cosmetic Act (FDCA)	EPA approves small-scale field trials and commercialization. Pesticides cannot have "unreasonable adverse effects on the environment"; as defined by FIFRA, this means "any unreasonable risk to man or the environment, taking into account the economic, social, and environmental costs and benefits of the use of any pesticide." This analysis is based on explicit risk–benefit tradeoffs; when any significant hazard is identified (which has been rare for biotechnology pesticides such as plant-incorporated protectants or microbial pesticide products), analyses include economic benefits and comparisons to current pesticides, how they are used, and their environmental effects. Post-market oversight is strong with monitoring and reporting requirements, and by statute, products must be reevaluated at least every 15 years; however, EPA has been reevaluating biotechnology products every 5–6 years. EPA also has the flexibility under FIFRA to reassess a registration decision at any time if new information suggests the probability of adverse effects may be greater than what was originally estimated.
EPA	Intergeneric microorganism	Toxic Substances Control Act (TSCA)	For field trials, product developers submit a TSCA Environmental Release Application for approval. The product cannot have "unreasonable risk of injury to human health or the environment." This analysis is based on risk–benefit tradeoffs. If some hazard is identified (which has been rare in EPA's experience as of 2016), then the risks relating to that hazard are quantified and compared with expected benefits. Such benefits can include economic analyses and environmental benefits accrued from replacing previous generations of technology. EPA is time-limited to 60 days for its analysis, but product developers often agree to a time extension to gather necessary data. For commercial use, a product developer submits a Microbial Commercial Activity Notification. EPA then must find that the product has "unreasonable risk" or the product will move forward after 90 days. Again, product developers often agree to a time extension. Once a microorganism is in commerce, the manufacturer is required to report any adverse-effect information through Section 8(e) of TSCA.[a]

continued

TABLE 3-3 Continued

Agency	Product	Statute	Authority
EPA	New chemical (including some RNAi)	TSCA	For commercial use, a product developer submits a Pre-Manufacture Notice. EPA then must find that the product has "unreasonable risk" or the product will go to market after 90 days. If environmental exposure is likely, EPA has certain data requirements that must be met.
			Once a chemical is in commerce, the manufacturer is required to report any adverse-effect information through Section 8(e) of TSCA.[a]
USDA–APHIS	Plant pest or regulated article (GE plant)	Plant Protection Act	Products are regulated by USDA–APHIS while under experimentation and can subsequently be deregulated for unconfined release, which for many types of products is a necessary practical step for commercial use. Permitting for a field trial requires that the organism is adequately confined in transport or field environments so that it is unlikely to spread or cross with native species and does not pose concerns for threatened or endangered species. For pharmaceutical-producing crops, USDA–APHIS has explicit guidance for permitted field trials.
			A product developer submits a petition for deregulation, providing evidence that the article shows no more of a plant-pest risk than an equivalent non-GE organism; it needs to be unlikely to pose a plant-pest risk. This analysis is based only on risk.
			Once a product has been deregulated, there is virtually no post-market monitoring or oversight. However, some products remain under permit even when commercialized (e.g., plants that produce pharmaceuticals) and in those cases, USDA–APHIS maintains oversight.

[a]Reporting a TSCA Chemical Substance Risk Notice. Available at https://www.epa.gov/assessing-and-managing-chemicals-under-tsca/reporting-tsca-chemical-substantial-risk-notice. Accessed September 14, 2016.
SOURCE: Based on a white paper prepared for the committee by S. Carter, Science Policy Consulting, 2016, which is available upon request from the National Academies' Public Access Records Office at PARO@nas.edu.

the same genus. Examples would be algae engineered to produce biofuels or microorganisms engineered to extract metal from ore. TSCA grants EPA the authority to test and assess potential risks of the microorganism before it is brought to market. At the time the committee was writing its report, EPA had reached agreements with all product developers for regulated products the use or marketing of which the agency decided to restrict in some way. EPA has the authority under the amended TSCA to reevaluate a previously approved product; however, it is not clear how high biotechnology products will be prioritized compared to other chemicals for which EPA may wish to conduct reevaluation.

U.S. Department of Agriculture

USDA's Animal and Plant Health Inspection Service (APHIS) regulates biotechnology products under the Plant Protection Act (PPA), which protects plants and plant products from plant pests and noxious weeds, and under the Animal Health Protection Act (AHPA), which regulates products that are pests to or could cause disease in livestock. Less relevant to this discussion is that veterinary biologics are subject to regulation by USDA rather than by FDA under the Virus-Serum-Toxin Act.

With regard to the Plant Protection Act, USDA–APHIS oversight extends to those biotechnology plants that have been genetically engineered using a donor organism, a recipient organism, or a vector or vector agent that is listed in 7 C.F.R. Part 340 and meets the definition of a plant pest. The agency's oversight also applies to an unclassified organism or one whose classification is unknown. Oversight extends to products that contain an organism produced through genetic engineering via the use of a plant pest.[44]

Under the Animal Health Protection Act (7 U.S.C. § 8302), USDA–APHIS has oversight authority for biotechnology products that could directly or indirectly cause disease or damage to livestock. Its application is limited to pests that impact livestock (defined as farm-raised animals, including horses, cattle, bison, sheep, goats, swine, cervids, poultry and others, and farm-raised fish) and not to wildlife or fish that are not farm raised. Dissemination of livestock pest products or animal vectors that carry them can be restricted by limiting movement across state lines or import into the United States. USDA's assessment of whether a product poses a risk to livestock is triggered by processes such as the application for permit for interstate movement of a product. Genetically engineered insects are not treated differently than other insects under this authority; rather, the assessment is based solely on whether the insect carries any contagious, infectious, or communicable disease of livestock.

As of 2016, USDA authority over GE insects that are plant pests has been exercised under the PPA, but the AHPA authority has not been exercised over GE insects that are animal pests. USDA has reviewed and approved field trials of diamondback moth genetically engineered with population suppression genes under the PPA. FDA has also reviewed and approved field trials for mosquitoes (*Aedes aegypti*) genetically engineered for population suppression under the FDCA.[45]

Jurisdictional Gaps and Redundancies

Jurisdictional gaps are one of the major challenges that future biotechnology products pose to the existing scheme of environmental regulation under FIFRA, TSCA, PPA, and AHPA. Gaps occur if a biotechnology organism falls outside the definitions of products that can be regulated under those statutes and, therefore, does not receive premarket or post-market oversight and fails to receive environmental assessment (Box 3-3).[46]

Because the use of a plant pest to transform an organism has been a key feature of the regulations used by USDA–APHIS since 1987 to oversee products of biotechnology, the agency typically has not regulated biotechnology plants that are not engineered using a plant-pest vector or those plants that do not contain any plant-pest DNA. This potential gap in jurisdiction was identified as far back as 2000 in a National Research Council report (NRC, 2000) and has been discussed in the literature (see, for example, Kuzma and Kokotovich, 2011). Since 2011, USDA–APHIS has seen an increasing number of GE crops, plants, and plant products that do not fall within its jurisdiction (Camacho et al., 2014; Kuzma, 2016).

To provide clarity for product developers on whether a product is considered a regulated article by USDA–APHIS, the agency has begun soliciting and answering letters of inquiry from product developers on a case-by-case basis. The letter from the product developer contains information

---

[44]7 C.F.R. § 340.1.

[45]In January 2017, FDA issued a draft guidance to clarify that its definition of nonfood regulated articles no longer included those "intended to function as pesticides by preventing, destroying, repelling, or mitigating mosquitoes for population control purposes. FDA believes that this interpretation is consistent with congressional intent and provides a rational approach for dividing responsibilities between FDA and EPA in regulating mosquito-related products" (FDA, 2017b:6575).

[46]Many of these points were described previously by Carter et al. (2014).

BOX 3-3
The Glowing Plant: A Product Without a Regulatory Hook

If a biotechnology product does not contain a plant pest or a pesticide, is not a food or animal, is not an intergeneric microorganism in commerce, and does not produce a new chemical, who regulates it? This conundrum was presented to the regulatory agencies when the DIYbio Glowing Plant project in 2013 raised $484,013 in a crowdfunded Kickstarter campaign to create a bioluminescent plant using a non–plant pest engineering approach. Because it met none of the regulatory criteria, it could be available for distribution directly to consumers without regulatory review.

However, there was a regulatory twist. The Glowing Plant project was conceived as an open-source technology, and its developers encouraged others to further modify the plant genome using an available kit that makes use of plant-pest components for genome modification. If kit recipients used the kit-supplied *Agrobacterium* plant-pest components to further modify the plant genome, regulation by USDA–APHIS would be triggered, which would mean that the secondary developer would need to interact with USDA–APHIS to determine what regulations would apply to the specific situation. Furthermore, shipping of GE glowing-plant seeds across state lines could require a downstream notification to EPA under TSCA. Following concerns regarding the distribution of kits that contain GE components, Kickstarter ceased supporting projects that involve DIYbio kits, but other crowdfunding platforms continue to support them.

---

on the host organism, the intended phenotype, whether there is an intention to move or release the product, the intended genetic change in the final product (for example, insertion, deletion, or substitution of genetic material), a description of the vector or vector agent used to induce genetic change in the organism (for example, biolistic delivery, *Agrobacterium*-mediated, or site-specific nuclease), the identity of the gene construct, and information on all elements of the gene construct (type, name, source, and function). As of 2016, USDA–APHIS had considered several cases of crops engineered with genome-editing technology to cause directed insertions or deletions of one to several bases. Several instances involving directed deletions or insertions have been deemed not regulated under 7 C.F.R. Part 340.[47] USDA–APHIS has made similar decisions to not regulate GE crops and plants engineered through biolistics if the resulting crop does not contain plant-pest sequences. However, some genome-edited crops have been regulated if they incorporated insertions with plant-pest sequences. USDA–APHIS responses to letters of inquiry often contain language with recommended actions for developers to ensure that plant-pest genetic sequences will not be present in the final product. In these situations, USDA has not required premarket testing, reported a risk finding, or undertaken a NEPA analysis.

In January 2017, USDA–APHIS issued a proposed rule (USDA–APHIS, 2017), which clarified that GE organisms in which the sole modification was a deletion or a single base-pair substitution that could otherwise be obtained through mutagenesis would not fall under the agency's regulatory purview. A modification that introduces only naturally occurring sequences from sexually compatible relatives that could be achieved through conventional-breeding methods also would fall outside USDA–APHIS's oversight. The same would be true for null segregants, that is, the progeny of organisms with introduced DNA that have not inherited the introduced sequences. Organisms that contained recombinant DNA or synthetic DNA would be subject to oversight. The proposed rule was open for public comment at the time the committee's report was published.

---

[47]See Submission Process for Am I Regulated Letters of Inquiry, https://www.aphis.usda.gov/aphis/ourfocus/biotechnology/am-i-regulated/regulated+article+letters+of+inquiry/regulated+article+letters+of+inquiry. Accessed October 11, 2016.

Even when products fall within an agency's jurisdiction, the product still may not receive pre-market or post-market oversight. The sponsors of GE crops and plants regulated by USDA–APHIS can petition the agency for nonregulated status, which allows for commercialization. At the time the committee was writing its report, once a GE crop or plant is deregulated, the agency has virtually no monitoring or reporting requirements. However, some GE crops or plants that are not regulated by USDA–APHIS may fall under other authorities, including EPA's FIFRA (if the crop or plant contains a plant-incorporated protectant such as *Bacillus thuringiensis*). Under FIFRA, these crops or plants may be subject to post-market monitoring as a condition of registration; at the time the committee wrote its report, EPA required insect resistance management plans to delay the development of resistance by insect pests to the plant-incorporated protectant.

With regard to microorganisms altered or created by biotechnology, gaps in regulation exist depending on how the product is commercially used and what the genetic composition is. EPA's authority over such microorganisms applies to those that are intergeneric and are sold or distributed; products produced and only used within the home may not be regulated under TSCA.

One way to address these sorts of gaps is for EPA to exercise authority under TSCA. In the 1986 Coordinated Framework, TSCA was described as a "back-stop" authority that could be applied to any biotechnology organism that did not fall under other authorities. EPA's authority under TSCA over a variety of biotechnology organisms, including plants and animals, was reaffirmed in 2001 with the Executive Office of the President's Office of Science and Technology Policy and Council on Environmental Quality's biotechnology case studies (CEQ and OSTP, 2001). However, at the time the committee was writing its report, EPA had not exercised its authority under TSCA in this way; whether this is due to insufficient resources or interpretations of authority is not clear. It would be helpful if the federal government would make a policy determination as to whether EPA will serve this gap-filling role.

Another way to address the gap is through the rulemaking process to update regulations. USDA–APHIS began such a process in early 2016 to address the gap pertaining to GE crops and plants. USDA published a Notice of Intent (NOI) to perform a programmatic environmental impact statement to capture not only GE crops and plants posing plant-pest risk but also those that pose potential noxious-weed risks. The definition of a noxious weed broadly covers potential harms such as "damage . . . to the natural resources of the United States, the public health, or the environment."[48] At the time the committee was writing its report, USDA–APHIS had routinely interpreted these authorities in a limited way to plants that are aggressively invasive, have significant negative impacts, and are extremely difficult to manage or control once established. Under the NOI, USDA–APHIS proposed four options (USDA–APHIS, 2016a):

1. Take no action and continue to regulate GE organisms as it historically has done.
2. Revise regulations to implement a two-step process that would first analyze the potential of a product of biotechnology to pose plant health risks and then determine the use of any regulatory action as appropriate or needed.
3. Revise regulations to not only regulate products of biotechnology developed using a plant pest but also to regulate products of biotechnology that pose a risk as noxious weeds.
4. Cease regulation and implement a voluntary, nonregulatory consultative service for developers.

USDA–APHIS solicited public comments on the four options over the course of about 3 months to inform the scope of analysis in its subsequent environmental impact statement. The proposed rule issued in January 2017 followed the third option: regulation of products of biotechnology on the

---

[48]7 C.F.R. § 360.100.

basis of the use of a plant pest or on the basis of the product's noxious-weed risk (USDA–APHIS, 2017). Under this proposal, GE crops and plants would be subject to a mandatory regulatory status evaluation by USDA–APHIS only if the agency had not evaluated previously the plant-pest risk or noxious-week risk posed by the submitted trait–crop combination or if the trait–crop combination has received DNA from a donor organism in a taxon known to contain plant pests and the introduced DNA was sufficient to produce a plant-disease property in the trait–crop combination. The agency would also have the ability to reevaluate its decision on a deregulated article if new information became available that indicated the organism may cause a plant or noxious-weed risk. The proposed rule was open for public comment at the time the committee's report was published.

An additional challenge with regard to environmental protection is that agencies may be able to identify environmental effects in their assessments of the products but lack the jurisdictional authority to address these effects. FDA and USDA, like most other federal agencies, are responsible for assessing environmental effects of its major federal actions under NEPA (Box 3-4) and must

---

BOX 3-4
The Role of the National Environmental Policy Act in Environmental Protection

The National Environmental Policy Act (NEPA) imposes a procedural requirement for federal agencies to consider the environmental effects of their major federal actions by conducting an environmental assessment, preparing an environmental impact statement, or determining that the action fits within a categorical exclusion. USDA–APHIS and FDA use NEPA to extend their regulatory reach to better address environmental concerns. (EPA's assessments under TSCA and FIFRA are considered to be equivalent to a NEPA assessment and so actions under those statutes are exempt.) The NEPA process and its equivalents also permit explicit inclusion of economic and market effects and provide for public participation and comment.

It is important to note that even when NEPA is triggered and an agency prepares an environmental assessment or an environmental impact statement, the agency is not granted any additional authority to regulate based on the outcome of those evaluations. For example, USDA–APHIS can still consider only plant health in its regulatory decision making.[a] For FDA, this predicament has been controversial when the agency has regulated biotechnology-altered animals under its new animal drug regulations for safety and effectiveness endpoints for the animal but does not have authority to manage the environmental risks such products may pose (Earthjustice and CFS, 2013). However, the experience arising from legal action (*Center for Food Safety v. Vilsack*, 2013) to move agencies toward use of the NEPA trigger has resulted in a stronger assessment process. For instance, USDA–APHIS has shown greater emphasis on the use of environmental assessments as a means to avoid the more costly and time-consuming environmental impact statements, and this is reflected in more comprehensive risk assessments even though decisions are still ultimately focused on plant-pest considerations.

For products that fall under USDA–APHIS's purview under the Plant Protection Act, significant federal actions that may trigger NEPA include permitting[b] and deregulation of products. When NEPA is triggered, USDA–APHIS must determine its appropriate response under NEPA regulations before a permit can be granted or the product deregulated. This assessment includes any likely environmental risks and requires a comparative analysis of other possible actions, including no action. USDA–APHIS evaluates the environmental assessment and usually is able to issue a Finding of No Significant Impact (FONSI). If the environmental assessment shows some significant risk, then an environmental impact statement must be developed. An environmental impact statement is a much more comprehensive document, often taking years to develop. Once a FONSI has been issued or an environmental impact statement has been completed, USDA–APHIS makes its regulatory decision based on its own statutory authorities (that is, on a product's potential plant-pest risks). The NEPA process provides a strong incentive for product developers to ensure minimal risks so that they can receive a FONSI and avoid the time and expense required for a full environmental impact statement.

BOX 3-4 Continued

USDA–APHIS conducted only environmental assessments for GE crops until it was ordered by a federal court in 2010 to conduct environmental impact statements for GE alfalfa and sugar beet (Cowan and Alexander, 2013). Between 2013 and September 2016, USDA–APHIS required a full environmental impact statement for deregulation decisions on five GE crops and plants out of 25 total (20 percent of the time).[c] Although much of the agency's experience is with crops, it has also regulated GE insects that are plant pests, including the GE pink bollworm, medfly, and diamondback moth. The pink bollworm and medfly underwent a full programmatic environmental impact statement for a field trial in 2008 (USDA–APHIS, 2008). The diamondback moth was issued a FONSI in 2014 for a field trial (USDA–APHIS, 2014), but before open field trials began in 2016 the FONSI was withdrawn because the agency had failed to formally advise the public about the trial in a second notice; as of November 2016, the agency was preparing an environmental assessment for a new permit application for field trials of the moth (USDA–APHIS, 2016b).

For FDA's oversight of biotechnology-altered animals, NEPA can be triggered for major federal actions, which include FDA's approval of an investigational new animal drug application[d] (for example, an application to commence a field trial) and FDA's approval of a new animal drug application to commence for commercial use. The Oxitec mosquitoes (engineered to produce sterile offspring) were the first biotechnology-altered animals to undergo a NEPA assessment for a field release. Oxitec's draft environmental assessment (Oxitec, 2016) and FDA's preliminary FONSI (FDA, 2016a) were released for public comment in March 2016, and the mosquitoes were approved for release by FDA in August 2016 when the agency published a final environmental assessment and a final FONSI (FDA, 2016b,c). Local authorities in the Florida county where the mosquito trial had been proposed placed two nonbinding referendums on voters' ballots in the November 2016 election on whether to approve having the mosquito trial take place in their jurisdiction. One referendum was for the residents of Key Haven, where the trials were to be conducted; the other was for the residents of the county. Voters in Key Haven voted their referendum down while the majority of county voters approved the measure to allow the mosquito trial to proceed. At the time the committee was writing its report, the staff of the area mosquito control district was working with Oxitec to identify a location for the trial (Allen, 2016).

As of September 2016, FDA had issued FONSIs for new animal drug applications for several GE animals under confinement conditions, thus clearing them for use. These include a goat that produces a pharmaceutical in its milk, growth-enhanced salmon, and a chicken that produces a pharmaceutical in its eggs. FDA has also cleared a GE mosquito for field trials under an investigational new animal drug application. It reviewed a zebra fish genetically engineered to fluoresce before the agency put in place its GE animal policy in 2009; the agency decided not to enforce the approval requirement for zebra fish based on an evaluation of risk factors, including whether it posed a human, animal, or environmental risk.

In 2015, the agency approved the AquAdvantage® salmon from the company AquaBounty Technologies. The GE salmon was engineered with a growth hormone promoter to grow to market size twice as fast as wild-type salmon (FDA, 2015b,c). Although FDA does not have the authority to regulate based on environmental endpoints, it negotiated with AquaBounty to adopt practices that would mitigate potential risks. Not only did these practices allow FDA to issue a FONSI and avoid an environmental impact statement, but they became part of the application itself. If AquaBounty does not adhere to the practices in the approved application, FDA can revoke the approval.[e]

---

[a]A court recently confirmed that USDA–APHIS does not have the legal authority to regulate based on environmental impacts that may be revealed by a NEPA assessment beyond those related to plant health: *Center for Food Safety v. Vilsack*, 2013, 718 F.3d 829 (9th Cir. May 17).

[b]Some low-risk products are exempt from permitting procedures and instead use an expedited notification. This notification process does not trigger NEPA.

[c]See Petitions for Determination of Nonregulated Status at https://www.aphis.usda.gov/biotechnology/petitions_table_pending.shtml. Accessed September 30, 2016.

[d]For drugs and animal drugs, FDA's regulatory oversight begins when the product developer opens an investigational new drug (or investigational new animal drug) file with FDA, which usually occurs when the drug is ready for clinical trials. For GE animals, FDA encourages product developers to talk to agency representatives as early as possible.

[e]Note the requirements in FDA's approval letter and the appendix that specifies specific measures (FDA, 2015d).

SOURCE: Based on a white paper prepared for the committee by S. Carter, Science Policy Consulting, 2016, which is available upon request from the National Academies' Public Access Records Office at PARO@nas.edu.

BOX 3-5
The Role of the Endangered Species Act in Environmental Protection

As biotechnology products become more widely used in open environments, they may have impacts on endangered species. In addition to requirements for the federal agencies, the Endangered Species Act (ESA) may also apply to companies and private individuals; the full extent of the ESA's impact on biotechnology-product deployment remains an open issue. Federal action triggers the ESA, which is administered by the U.S. Fish and Wildlife Service (FWS) and the National Marine Fisheries Service (NMFS). Agencies are required to assess the direct and indirect effects their action will have on endangered species and their critical habitat. At the time the committee was writing its report, analyses for most biotechnology products had resulted in a finding of "no effect" and so consultations with FWS and NMFS had not been required. If a federal agency concludes there may be an effect, then the agency must consult with the appropriate service (FWS, NMFS, or both) to determine if the action is "not likely" or "likely" to adversely affect the species. If the latter, then, as appropriate, FWS or NMFS will undertake an assessment to determine whether or not the action will cause jeopardy and, if so, what reasonable prudent alternatives and measures the agency could employ to preclude jeopardy and potentially allow a specified level of adverse effects on the species of interest.

The ESA process has been controversial, particularly with respect to pesticides and EPA, with an inability of EPA to find a common approach with FWS and NMFS to assess risks to endangered species. Although some efforts have attempted to find a solution to this problem,[a] there is a fundamental difference in the risk assessments that biotechnology organisms face in their product-based regulation (which can accommodate some level of reasonable risk) and those that would be conducted under the ESA (which attempt to determine harm to even a single individual of a species). FWS and NMFS can and sometimes do authorize some level of adverse effects (also known as "take"), but they err on the side of the endangered species.

With respect to biotechnology products, the regulatory agencies have limited experience undertaking ESA Section 7(a)(2) consultations. At the time the committee was writing its report, EPA, FDA, and USDA–APHIS have typically made "no effect" determinations for their regulatory actions. However, at the time the committee was writing its report, FDA was being sued for its approval of the AquAdvantage salmon for inadequate compliance with the ESA (along with other complaints).[b] FDA briefed FWS and NMFS on the product and received letters of concurrence for its "no effect" finding, but this was the extent of the agencies' interactions. As of 2016, FWS and NMFS had had few opportunities to provide input; therefore,

---

consider impacts of their decisions on endangered species under the ESA (Box 3-5). However, they are restricted by their authorities from considering certain risk-assessment endpoints.

There is also the potential for jurisdictional redundancy; for instance, all agencies receive plant composition and agronomic performance data for GE crops, and these data are reviewed for various overlapping concerns in relation to risks to agriculture, health, and the environment. Findings could vary among the agencies based on differing assessment considerations. Neither the lack of oversight nor redundancy necessarily indicates a risk, but either may erode public confidence and confuse developers. The risks would depend on the products; at the time the committee was writing its report, most risks had been successfully managed under the Coordinated Framework on a case-by-case basis.

SUMMARY AND CONCLUSIONS

The Coordinated Framework for Regulation of Biotechnology is a complex collection of statutes and regulations that provide the basis for federal oversight of biotechnology products. The Coordinated Framework appears to have considerable flexibility to cover a wide range of biotechnology products, although in some cases the agencies' jurisdiction has been defined in ways that

these agencies may not be overly familiar with biotechnology and emerging biotechnology issues that will be important when considering future biotechnology products.

Several scientists have raised the question why other agencies, particularly FWS and NMFS, are not involved in field releases of biotechnology-altered animals, given their expertise in ecology and their administration of the ESA (Balint, 1999; Kelso, 2004; Logar and Pollock, 2005; Otts, 2014). In addition to its authority under the ESA, FWS also has authority under the Lacey Act (18 U.S.C. § 42) to prohibit the importation and transportation of species "injurious to human beings, to the interests of agriculture, horticulture, forestry, or to wildlife or the wildlife resources of the US." It could be argued that on the basis of the Lacey Act, FWS had authority over GE mosquitoes such as the one developed by Oxitec (see Box 3-4). However, FWS has not been intimately involved in the regulation of GE animals, and FWS personnel have criticized FDA's environmental assessments and decision-making processes for AquAdvantage Salmon (as summarized in Earthjustice and CFS, 2013).

The ESA's authorities to prohibit "take" of listed species, including significant modification or degradation of habitat, continue after a product is deployed. For unregulated products (likely including those that have been reviewed and deregulated by USDA–APHIS), the ESA can be enforced against the company or even the individual that deploys the product if it results in some take of endangered species. However, if a product has been approved or permitted by a federal agency and a product developer follows the terms of that approval or permit, then the product developer may be shielded from enforcement action. Any unintended harm to endangered species would instead trigger a new consultation between the federal agency and FWS and NMFS. In this case, if FWS and NMFS were to find that the product is likely to adversely affect a listed species, then the agency may be required to adjust the terms of (or revoke) the approval or permit.[c]

---

[a]The National Research Council released a report trying to find common ground in 2013, *Assessing Risks to Endangered and Threatened Species from Pesticides* (NRC, 2013).

[b]*Institute for Fisheries Resources et al. v. Burwell et al.,* case number 3:16-cv-01574, in the U.S. District Court for the Northern District of California.

[c]Although this situation has not come up in the context of biotechnology products, an example has arisen for biological control agents, which are permitted by USDA–APHIS under the Plant Protection Act. A permit for a leaf beetle deployed to control saltcedar, an invasive weed, was terminated due to concerns about the habitat of an endangered bird, the southwestern willow flycatcher. See https://www.aphis.usda.gov/aphis/ourfocus/planthealth/plant-pest-and-disease-programs/sa_environmental_assessments/ct_saltcedar. Accessed September 30, 2016.

SOURCE: Based on a white paper prepared for the committee by S. Carter, Science Policy Consulting, 2016, which is available upon request from the National Academies' Public Access Records Office at PARO@nas.edu.

potentially may leave gaps or overlaps in regulatory oversight. Even when the statutes technically do allow agencies to regulate these products, the current statutes equip the regulators with tools that may, at times, make it hard for them to regulate the products effectively. For example, the statutes may not empower regulators to require product sponsors to share in the burden of generating information about product safety, may place the burden of proof on regulators to demonstrate that a product is unsafe before they can take action to protect the public, or may require cumbersome processes or procedures the regulators must follow before they can act, and almost all of the statutes lack adequate legal authority for post-marketing surveillance, monitoring, and continuous learning approaches. Thus, although the products of future biotechnology are often likely to be within the jurisdiction of existing regulators, they may struggle to regulate these products effectively and to respond nimbly to the products that will be coming. Clearly gaps in the process will emerge as novel products are brought forward in future years, but they cannot be anticipated in any regulatory system, so they must be addressed as they emerge within the system.

The diversity of open-release products under development continues to expand, as described in Chapter 2, but it is not clear how many new products per year are likely to be submitted to federal agencies for premarket review or post-market oversight during the next decade. It is, however, reasonable to assume the number of products per year that will require federal oversight will increase and the complexity of future assessments for these products, and the associated level of effort required on the part of appropriate regulatory authorities, will also increase.

Conclusion 3-1: The diversity of biotechnology consumer products anticipated over the next decade confronts consumer- and occupational-safety regulators with two related challenges: (1) to find jurisdiction under existing statutes to regulate all the products that may pose risks to consumers and (2) to utilize the risk-assessment tools available under those statutes to provide oversight that protects consumers while allowing beneficial innovation.

Existing statutes offer promising pathways to meet these challenges, although there may be cases when a novel product falls outside the jurisdiction of EPA, FDA, or USDA and is either in a jurisdictional gap (where no regulator has authority to address potential safety concerns) or under the jurisdiction of another agency, such as CPSC, that has fewer statutory authorities and capabilities to conduct rigorous, timely risk assessment. For this reason, EPA, FDA, and USDA may at times need to make use of the flexibility available under their statutes to minimize gaps in jurisdiction and to position novel products under the statutory framework most suited to each product's characteristics and level of risk. Specifically, a federal policy determination as to whether EPA will exert a gap-filling role under TSCA would reduce uncertainty.

Even in cases when EPA, FDA, and USDA interpret their jurisdiction as widely as courts will allow them to do, it is still true that a growing number of future biotechnology products may fall into gaps in these agencies' jurisdiction. For example, some biotechnology products, such as toys or domestic products manufactured in the home for domestic use, may not be covered by EPA, FDA, or USDA and may fall mainly under CPSC's jurisdiction, while other biotechnology products used in the workplace as means of industrial production may fall mainly under OSHA's jurisdiction.

Conclusion 3-2: The Consumer Product Safety Act and the Occupational Safety and Health Act do not provide CPSC and OSHA with legal authorities and tools that are well tailored to the challenges of regulating novel biotechnology-based consumer products and means of production.

CPSC lacks authority to conduct premarket safety analysis of biotechnology-based consumer products and has limited tools for responding to risks emerging after products are marketed. The Occupational Safety and Health Act has limitations that may make it difficult for OSHA to respond nimbly to novel uses of biotechnology as a means of production.

Conclusion 3-3: The shifting structure and financing of biotechnology-product development and manufacturing may place strains on existing systems for research oversight, timely response to premarket notifications, and post-marketing inspections and product testing.

Appropriate staffing of the regulatory agencies is critical, as is careful planning to maximize the reach of available research regulations (INAD, IDE, IND, and EUP) for research not subject to the NIH guidelines. Federal oversight of biotechnology products can take place at a number of phases in the product-development cycle, depending on the product and its intended use. Under current statutes, regulators' power to require product manufacturers to submit premarket safety studies and post-marketing safety information varies, depending on the product type.

Conclusion 3-4: Under existing statutes, much of the burden to generate evidence of consumer safety falls on regulators and public funding agencies; this implies that adequate federal support for research will be crucial to protect consumer and occupation safety.

Under many of the statutes that the committee reviewed, regulators have only limited authority to require product sponsors to conduct (and thus to fund) studies to generate information about the safety of their products, and some statutes require regulators to bear the burden of establishing that a human safety or environmental problem exists before the regulator can take steps to manage the risk. Such policies foster innovation by reducing the barriers to entry and costs of bringing new products onto the market, but they imply a corresponding obligation for regulators and public funding agencies to generate safety information through inspections, product testing, information gathering, and research to identify and analyze potential risks. These policies reflect a balance Congress has struck between the benefits of innovation and the burdens of assuming major federal responsibilities to generate information that is necessary to protect the public and the environment. As a profusion of diverse biotechnology products enters the market during the next 5–10 years, maintaining this balance will require adequate federal support for the research, inspection, testing, and other activities that Congress has confided to federal government agencies under the existing statutes. There are examples, however, that suggest that the federal investment in information-gathering activities can be leveraged through public–private partnerships, public engagement, and other measures that mobilize data and know-how already existing within industry and the product-consuming public.

Conclusion 3-5: Post-market risk identification, analysis, and safety surveillance are important tools for supporting beneficial uses of innovative products and ensuring public and environmental safety.

For the product categories covered in this report, existing statutes do not provide regulators with a complete set of modern tools, such as authorization to develop active surveillance systems and shared data resources to support regulatory science and continuous learning. Therefore, publicly funded research has an ongoing role to play throughout the entire product life cycle, although public–private partnerships offer promise for both mobilizing know-how and leveraging public investments.

Conclusion 3-6: For some product categories considered in this report, consumer-safety regulatory agencies have little or no authority to restrict the receipt, use, sale, or distribution of products to address risks that otherwise-safe products may pose in the hands of unqualified or malicious users.

Existing programs to restrict shipments of biological agents (the Federal Select Agent Program administered by USDA–APHIS and CDC, U.S. Department of Commerce restrictions, and U.S. Department of Health and Human Services voluntary screening programs) are neither focused nor scaled to address concerns that may arise in connection with future consumer products.

The future industry structure will include new players, such as do-it-yourself and at-home biotechnology enthusiasts and product developers, nontraditional manufacturers, and those entering the biotechnology space with support from nontraditional funding sources for research. Protecting public safety may, at times, call for controls over who can access and use certain types of products—for example, to restrict the use of the product to qualified users or to ensure the product is used only in facilities that agree to implement certain safety measures. The regulatory agencies of the Coordinated Framework, such as FDA, OSHA, and CPSC, have little authority to restrict sales, distribution, and use of products. The federal frameworks currently in place for limiting access to biological agents (the Federal Select Agent Program administered by USDA–APHIS and CDC, U.S. Department of Commerce restrictions on transactions, and voluntary screening programs administered by the U.S. Department of Health and Human Services) are geared toward controlling small numbers of highly dangerous or strategically significant products, rather than a wider array of biotechnology products that may require qualified users in order to be safe.

> **Conclusion 3-7: Definitions of what constitutes a product of biotechnology under different statutes lead to jurisdictional gaps and redundancies in the assessment of the environmental effects of biotechnology products.**

EPA, FDA, and USDA may not have authority to conduct an environment risk assessment of a future biotechnology product if the product does not fall within the parameters of their authorizing statutes. Even when FDA and USDA identify environmental effects through the NEPA process or through ESA consultations with the U.S. Fish and Wildlife Service or the National Marine Fisheries Service, they do not have the authority to make regulatory decisions based on the findings of the environmental assessment or environmental impact statement. As of 2016, most existing biotechnology products have been used in contained environments or open environments in managed systems, but with more biotechnology products designed for released into open environments with minimal or no management, this disconnect may be more acute.

Jurisdiction redundancy also exists; for example, all three agencies can receive composition and agronomic performance data for GE crops. Each agency has a different approach to assessing risks related to agriculture, health, and the environment; therefore, the findings of each agency may differ slightly from the others and may create confusion for product developers.

REFERENCES

Allen, G. November 20, 2016. Florida Keys approves trial of genetically modified mosquitoes to fight Zika. National Public Radio. Available at http://www.npr.org/sections/health-shots/2016/11/20/502717253/florida-keys-approves-trial-of-genetically-modified-mosquitoes-to-fight-zika. Accessed December 22, 2016.

Balint, P.J. 1999. Marine biotechnology: A proposal for regulatory reform. Politics and the Life Sciences 18(1):25–30.

Bartis, J.T., and E. Landtree. 2006. Nanomaterials in the Workplace: Policy and Planning Workshop on Occupational Safety and Health. Santa Monica, CA: RAND Corporation. Available at http://www.rand.org/content/dam/rand/pubs/conf_proceedings/2006/RAND_CF227.pdf. Accessed September 30, 2016.

Camacho, A., A. Van Deynze, C. Chi-Ham, and A.B. Bennett. 2014. Genetically engineered crops that fly under the US regulatory radar. Nature Biotechnology 32:1087–1091.

Carter, S.R., M. Rodemeyer, M.S. Garfinkel, and R.M. Friedman. 2014. Synthetic Biology and the U.S. Biotechnology Regulatory System: Challenges and Options. J. Craig Venter Institute. Available at http://www.jcvi.org/cms/fileadmin/site/research/projects/synthetic-biology-and-the-us-regulatory-system/full-report.pdf. Accessed August 10, 2016.

CDC (Centers for Disease Control and Prevention). July 4, 2003. Nonfatal dog bite–related injuries treated in hospital emergency departments—United States, 2001. Morbidity and Mortality Weekly Report 52(26):605–610.

CEQ and OSTP (Council on Environmental Quality and Office of Science and Technology). 2001. CEQ and OSTP Assessment: Case Studies of Environmental Regulations for Biotechnology. Available at www.whitehouse.gov/files/documents/ostp/Issues/ceq_ostp_study1.pdf. Accessed September 30, 2016.

Charo, R.A. 2015. The legal/regulatory context. Pp. 13–19 in International Summit on Human Gene Editing: A Global Conversation, Commissioned Papers. Washington, DC: Chinese Academy of Sciences, Royal Society, U.S. National Academy of Sciences, and U.S. National Academy of Medicine. Available at http://nationalacademies.org/cs/groups/pgasite/documents/webpage/pga_170455.pdf. Accessed October 1, 2016.

Cowan, T., and K. Alexander. 2013. Deregulating Genetically Engineered Alfalfa and Sugar Beets: Legal and Administrative Responses. Washington, DC: Congressional Research Service.

Crawford, L.M. July 17, 2003. FDA's Role in the National Response to an Emerging Zoonotic or Secondarily Transmitted Infectious Disease. Statement of Lester M. Crawford, Deputy Commissioner, U.S. Food and Drug Administration, Before the Senate Committee on Environmental and Public Works, 108th Congress. Available at http://www.fda.gov/newsevents/testimony/ucm115100.htm. Accessed September 30, 2016.

Culleen, L.E. May 30, 2016. The TSCA amendments simplified: Nine key features of the new law and three compromises that will affect business. Bloomberg BNA Chemical Regulation Reporter. Available at http://www.arnoldporter.com/~/media/files/perspectives/publications/2016/05/the-tsca-amendments-simplified-nine-key-features-of-the-new-law-and-three.pdf. Accessed September 30, 2016.

Duvall, M.N., R.J. Carra, T.M. Serie, and S.A. Kettenmann. June 2, 2016. What's new about the revised TSCA—Toxic Substances Control Act. National Law Review. Available at http://www.natlawreview.com/article/what-s-new-about-revised-tsca-toxic-substances-control-act. Accessed September 30, 2016.

Earthjustice and CFS (Center for Food Safety). 2013. Comments Regarding the U.S. Food and Drug Administration's Draft Environmental Assessment and Preliminary Finding of No Significant Impact Concerning a Genetically Engineered Atlantic Salmon. Docket No. FDA-2011-N-0899. April 26. Available at http://www.centerforfoodsafety.org/files/42613-final-cfs-and-earthjustice-comment_49553.pdf. Accessed September 30, 2016.

Elliot, S.J. September 22, 2016. GMO food labeling and CRISPR. Online. PharmaPatents blog. Available at https://www.pharmapatentsblog.com/2016/09/22/gmo-food-labeling-and-crispr. Accessed September 30, 2016.

EOP (Executive Office of the President). 2011. Executive Order 13563 of January 18, 2011: Improving Regulation and Regulatory Review. Federal Register 76:3821–3823.

EOP. 2016. National Strategy for Modernizing the Regulatory System for Biotechnology Products. Available at https://obamawhitehouse.archives.gov/sites/default/files/microsites/ostp/biotech_national_strategy_final.pdf. Accessed January 31, 2017.

EOP. 2017. Modernizing the Regulatory System for Biotechnology Products: An Update to the Coordinated Framework for the Regulation of Biotechnology. Available at https://obamawhitehouse.archives.gov/sites/default/files/microsites/ostp/2017_coordinated_framework_update.pdf. Accessed January 30, 2017.

EPA (U.S. Environmental Protection Agency). 1997. Microbial Products of Biotechnology; Final Regulation Under the Toxic Substances Control Act. Federal Register 62:17910–17958.

Evans, B.J. 2009. Seven pillars of a new evidentiary paradigm: The Food, Drug, and Cosmetic Act enters the genomic era. Notre Dame Law Review 85:419–524.

Evans, B.J. 2015a. Governance at the Institutional and National Level. Presentation at the International Summit on Human Gene Editing: A Global Discussion, sponsored by the Chinese Academy of Sciences, Royal Society, U.S. National Academy of Sciences, and U.S. National Academy of Medicine, December 2, Washington, DC.

Evans, B.J. 2015b. The limits of FDA's authority to regulate clinical research involving high-throughput DNA sequencing. Food and Drug Law Journal 70(2):259–287.

Fama, R. September 8, 2016. The new GMO labeling law: A matter of perspective. Food Safety News. Online. Available at http://www.foodsafetynews.com/2016/09/the-new-gmo-labeling-law-a-matter-of-perspective/#.V9_YZE0m7IU. Accessed September 30, 2016.

FDA (U.S. Food and Drug Administration). 1992. Statement of Policy: Foods Derived from New Plant Varieties. Federal Register 57:22984–23005.

FDA. 1997. Consultation Procedures under FDA's 1992 Statement of Policy—Foods Derived from New Plant Varieties. Revised edition. Available at http://www.fda.gov/Food/GuidanceRegulation/GuidanceDocumentsRegulatoryInformation/Biotechnology/ucm096126.htm. Accessed September 30, 2016.

FDA. 2003. Guidance for Industry: Interim Procedures for Qualified Health Claims in the Labeling of Conventional Human Food and Human Dietary Supplements. Available at http://www.fda.gov/Food/GuidanceRegulation/GuidanceDocumentsRegulatoryInformation/ucm053832.htm. Accessed September 30, 2016.

FDA. 2011. Guidance for Industry and FDA Staff: Classification of Products as Drugs and Devices & Additional Product Classification Issues. Draft Guidance. June. Available at http://www.fda.gov/downloads/RegulatoryInformation/Guidances/UCM258957.pdf. Accessed September 30, 2016.

FDA. 2015a. Guidance for Industry: Regulation of Genetically Engineered Animals Containing Heritable DNA Constructs. Available at http://www.fda.gov/downloads/animalveterinary/guidancecomplianceenforcementguidanceforindustry/ucm113903.pdf. Accessed September 30, 2016.

FDA. 2015b. AquAdvantage® Salmon Environmental Assessment. November 12. Available at http://www.fda.gov/downloads/AnimalVeterinary/DevelopmentApprovalProcess/GeneticEngineering/GeneticallyEngineeredAnimals/UCM466218.pdf. Accessed September 30, 2016.

FDA. 2015c. Finding of No Significant Impact: AquAdvantage Salmon. November 12. Available at http://www.fda.gov/downloads/AnimalVeterinary/DevelopmentApprovalProcess/GeneticEngineering/GeneticallyEngineeredAnimals/UCM466219.pdf. Accessed September 30, 2016.

FDA. 2015d. AquAdvantage Salmon Approval Letter and Appendix. Available at http://www.fda.gov/AnimalVeterinary/DevelopmentApprovalProcess/GeneticEngineering/GeneticallyEngineeredAnimals/ucm466214.htm. Accessed September 30, 2016.

FDA. 2016a. Preliminary Finding of No Significant Impact in Support of an Investigational Field Trial of OX513A *Aedes aegypti* Mosquitoes. March. Available at http://www.fda.gov/downloads/AnimalVeterinary/DevelopmentApprovalProcess/GeneticEngineering/GeneticallyEngineeredAnimals/UCM487379.pdf. Accessed September 30, 2016.

FDA. 2016b. Environment Assessment for Investigational Use of *Aedes aegypti* OX513A. August 5. Available at http://www.fda.gov/downloads/AnimalVeterinary/DevelopmentApprovalProcess/GeneticEngineering/GeneticallyEngineeredAnimals/UCM514698.pdf. Accessed September 30, 2016.

FDA. 2016c. Finding of No Significant Impact in Support of a Proposed Field Trial of Genetically Engineered (GE) Male *Aedes aegypti* Mosquitoes of the Line OX513A in Key Haven, Monroe County, Florida under an Investigational New Animal Drug Exemption. August 5. Available at http://www.fda.gov/downloads/AnimalVeterinary/Development ApprovalProcess/GeneticEngineering/GeneticallyEngineeredAnimals/UCM514699.pdf. Accessed September 30, 2016.

FDA. 2017a. Regulation of Intentionally Altered Genomic DNA in Animals; Draft Guidance for Industry; Availability. Federal Register 82:6561–6564.

FDA. 2017b. Regulation of Mosquito-Related Products; Draft Guidance for Industry; Availability. Federal Register 82:6574–6575.

Felcher, E.M. 2008. The Consumer Product Safety Commission and Nanotechnology. Washington, DC: Woodrow Wilson International Center for Scholars. Available at http://www.nanotechproject.org/process/assets/files/7033/pen14.pdf. Accessed September 30, 2016.

Fischhoff, B. 2015. The realities of risk-cost-benefit analysis. Science 350(6260):aaa6516-1–aaa6516-7.

Frischmann, B.M., M.J. Madison, and K.J. Strandburg, eds. 2014. Governing Knowledge Commons. New York: Oxford University Press.

Goldsmith, S., and W.D. Eggers. 2004. Governing by Network: The New Shape of the Public Sector. Washington, DC: Brookings Institution Press.

Henley, L. 2013. How to Put Together an IDE Application. Slide 12 in presentation for FDA's 2013 Clinical Investigator Training Course, November 14, College Park, MD. Available at http://www.fda.gov/downloads/Training/ClinicalInvestigatorTrainingCourse/UCM378680.pdf. Accessed September 30, 2016.

Hutt, P.B., R.A. Merrill, and L.A. Grossman. 2014. Food and Drug Law: Cases and Materials, 4th Ed. St. Paul, MN: Foundation Press.

IRGC (International Risk Governance Council). 2015. Guidelines for Emerging Risk Governance. Lausanne, Switzerland: IRGC.

Kaplan, S., and B.J. Garrick. 1981. On the quantitative definition of risk. Risk Analysis 1:11–27.

Kelso, D.D.T. 2004. Genetically engineered salmon, ecological risk, and environmental policy. Bulletin of Marine Science 74(3):509–528.

Kettl, D.F. 2002. The Transformation of Governance: Public Administration for the Twenty-First Century. Baltimore: Johns Hopkins University Press.

Kuzma, J. 2016. Rebooting the debate about genetic engineering. Nature 531(7593):165–167.

Kuzma, J., and J.C. Besley. 2008. Ethics of risk analysis and regulatory review: From bio- to nanotechnology. Nanoethics 2(2):149–162.

Kuzma, J., and A. Kokotovich. 2011. Renegotiating GM crop regulation. EMBO Reports 12(9):883–888.

Levitt, J.A. March 20, 2001. FDA's Progress with Dietary Supplements. Statement of Joseph A. Levitt, Director, Center for Food Safety and Applied Nutrition, U.S. Food and Drug Administration, Before the House Committee on Government Reform, 107th Congress. Available at http://www.fda.gov/NewsEvents/Testimony/ucm115229.htm. Accessed September 30, 2016.

Lin, A.C. 2013. Prometheus Reimagined: Technology, Environment, and Law in the Twenty-first Century. Ann Arbor: University of Michigan Press.

Logar, N., and L.K. Pollock. 2005. Transgenic fish: Is a new policy framework necessary for a new technology? Environmental Science & Policy 8(1):17–27.

Meghani, Z., and J. Kuzma. 2011. The "revolving door" between regulatory agencies and industry: A problem that requires reconceptualizing objectivity. Journal of Agricultural and Environmental Ethics 24(6):575–599.

Mendeloff, J.M. 1988. The Dilemma of Toxic Substance Regulation. Cambridge, MA: MIT Press.

NIH (National Institutes of Health). 2016. NIH Guidelines for Research Involving Recombinant or Synthetic Nucleic Acid Molecules (NIH Guidelines). April. Available at http://osp.od.nih.gov/sites/default/files/resources/NIH_Guidelines.pdf. Accessed September 29, 2016.

NRC (National Research Council). 1983. Risk Assessment in the Federal Government: Managing the Process. Washington, DC: National Academy Press.

NRC. 1996. Understanding Risk: Informing Decisions in a Democratic Society. Washington, DC: National Academy Press.

NRC. 2000. Genetially Modified Pest-Protected Plants: Science and Regulation. Washington, DC: National Academy Press.

NRC. 2009. Science and Decisions: Advancing Risk Assessment. Washington, DC: The National Academies Press.

NRC. 2013. Assessing Risks to Endangered and Threatened Species from Pesticides. Washington, DC: The National Academies Press.

OSTP (Office of Science and Technology Policy). 1986. Coordinated Framework for Regulation of Biotechnology. Federal Register 51:23302. Available at https://www.aphis.usda.gov/brs/fedregister/coordinated_framework.pdf. Accessed August 10, 2016.

OSTP. 1992. Exercise of Federal Oversight Within Scope of Statutory Authority: Planned Introductions of Biotechnology Products into the Environment. Federal Register 57:6753–6762. Available at https://www.whitehouse.gov/sites/default/files/microsites/ostp/57_fed_reg_6753__1992.pdf. Accessed October 1, 2016.

Otts, S.S. 2014. U.S. regulatory framework for genetic biocontrol of invasive fish. Biological Invasions 16(6):1289–1298.

Oxitec. 2016. Draft Environment Assessment for Investigational Use of *Aedes aegypti* OX513A. February. Oxfordshire, UK: Oxitec, Ltd. Available at http://www.fda.gov/downloads/AnimalVeterinary/DevelopmentApprovalProcess/GeneticEngineering/GeneticallyEngineeredAnimals/UCM487377.pdf. Accessed September 30, 2016.

Renn, O. 1992. Concepts of risk: A classification. Pp. 53–79 in Social Theories of Risk, S. Krimsky and D. Golding, eds. Westport, CT: Praeger.

Renn, O. 2005. Risk Governance: Towards an Integrative Approach. Geneva: IRGC.

Rosa, E.A., O. Renn, and A.M. McCright. 2013. The Risk Society Revisited: Social Theory and Governance. Philadelphia: Temple University Press.

Rothenberg, E., K. McTique, and B. Nicksin. June 8, 2016. Congress Passes Amendments to TSCA. O'Melveny Alerts & Publications. Online. Available at https://www.omm.com/resources/alerts-and-publications/alerts/congress-passes-amendments-to-tsca. Accessed September 30, 2016.

Salamon, L.M., ed. 2002. The Tools of Government: A Guide to the New Governance. New York: Oxford University Press.

Shapiro, S.A., and T.O. McGarity. 1989. Reorienting OSHA: Regulatory alternatives and legislative reform. Yale Journal on Regulation 6(1):Article 2.

Shrader-Frechette, K. 2007. Nanotoxicology and ethical considerations for informed consent. Nanoethics 1(1):47–56.

Sidley. June 9, 2016. Significant changes to TSCA will affect a broad range of companies that manufacture and use chemicals. Sidley Environmental Update. Available at http://www.sidley.com/news/2016-06-08_environmental_update. Accessed September 30, 2016.

Stirling, A. 2007. Risk, precaution and science: Towards a more constructive policy debate. EMBO Reports 8(4):309–315.

Thompson, P. 2007. Food Biotechnology in Ethical Perspective, 2nd Ed. Dordrecht, Netherlands: Springer.

TPG (Thompson Publishing Group, Inc.). 2005. Lepay: FDA to take closer look at investigator-initiated trial. Guide to Good Clinical Practice Newsletter 12 (January).

USDA–APHIS (U.S. Department of Agriculture–Animal and Plant Health Inspection Service). 2008. Use of Genetically Engineered Fruit Fly and Pink Bollworm in APHIS Plant Pest Control Programs: Final Environmental Impact Statement. Available at https://www.aphis.usda.gov/plant_health/ea/downloads/eis-gen-pbw-ff.pdf. Accessed September 30, 2016.

USDA–APHIS. 2014. National Environmental Policy Act Decision and Finding of No Significant Impact: Field Release of Genetically Engineered Diamondback Moth Strains OX4319L-Pxy, OX4319N-Pxy, and OX4767A-Pxy. Available at https://www.aphis.usda.gov/brs/aphisdocs/13_297102r_fonsi.pdf. Accessed September 30, 2016.

USDA–APHIS. 2016a. Environmental Impact Statement; Introduction of the Products of Biotechnology. Federal Register 81:6225–6229.

USDA–APHIS. 2016b. Withdrawal of an Environmental Assessment for the Field Release of Genetically Engineered Diamondback Moths. Federal Register 81:78567.

USDA–APHIS. 2017. Importation, Interstate Movement, and Environmental Release of Certain Genetically Engineered Organisms. Federal Register 82:7008–7039.

Yescombe, E.R. 2007. Public–Private Partnerships: Principles of Policy and Finance. London: Elsevier.

You, E. 2016. Safeguarding the Bioeconomy: Looking Ahead. Webinar presentation to the National Academies of Sciences, Engineering, and Medicine Committee on Future Biotechnology Products and Opportunities to Enhance Capabilities of the Biotechnology Regulatory System, July 21.

4

Understanding Risks Related to Future Biotechnology Products

There is a history of risk assessments and regulatory determinations for biotechnology prod- ucts through the Coordinated Framework. However, the scope, scale, complexity, and tempo of products to be developed in the next 5–10 years (outlined in Chapter 2) will likely be substantially different from the scope, scale, complexity, and tempo of products developed between the 1980s and 2016. Under these new conditions, regulators will have to assess whether the risks of future products are different from products that have come before and whether the risk-analysis approaches that have been used (outlined in Chapter 3) are sufficient. If those approaches need to be revised for future products, then regulators will need to have the appropriate scientific capabili- ties, tools, and expertise to support oversight of those products.

RISKS FROM FUTURE BIOTECHNOLOGY PRODUCTS: SIMILARITIES TO THE PAST AND GAPS GOING FORWARD

As discussed in Chapter 3 (see Box 3-1), risks are comprised of undesirable outcomes (what), the possibility of occurrence (how likely), and state of reality (ways the risk occurs in pathways) (Renn, 1992). *Risk-assessment endpoints* are societal, human health, or environmental values that need to be managed or protected (NASEM, 2016a). There can be many pathways by which those risk-assessment endpoints are reached, and risk assessments provide a quantitative or qualitative evaluation of the endpoints.

The committee's statement of task posed two questions related to risk:

1. Could future biotechnology products pose different types of risks relative to existing prod- ucts and organisms?
2. Are there areas in which risks or lack of risks of biotechnology products are well understood?

The first question in the committee's charge was interpreted as a request to reflect on the degree to which regulatory human health and ecological endpoints used in risk assessments for existing biotechnology products are likely to be similar to or different from the endpoints that would be selected when assessing risks for future biotechnology products. For ecological risk-assessment endpoints, the committee considered the potential similarities and differences between explicitly defined ecological entities and their attributes within the ecosystems possibly at risk (see EPA, 1998) for existing biotechnology products and future biotechnology products. For human health risk-assessment endpoints, the similarities and differences considered for existing and future products were those associated with responses of individuals at the subcellular, cellular, tissue, and individual levels of biological organization; such responses typically serve as endpoints within specified human subpopulations (for example, see NRC, 2007, 2009). To compare specific risk-assessment endpoints between existing and future products, the committee was also asked to evaluate whether the exposure and effect pathways under which endpoints can be expressed differed between the two. That is, will the ways humans or an environment may be exposed to or the degree to which they may be affected by a future biotechnology product differ from the ways exposure and effect occur for existing biotechnology products? The U.S. Environmental Protection Agency (EPA, 1998), Suter (2007), and a 2009 National Research Council report (NRC, 2009) discuss the need to specify the spatial and temporal dimensionality of an assessment during its problem-formulation phase. Consideration of dimensionality also incorporates concepts of toxicity (NRC, 2007) and adverse outcome pathways (Ankley et al., 2010), which link subcellular and cellular perturbations to potential adverse outcomes at the tissue, organ, individual, population, or community level of biological organization. The committee incorporated the perspectives on dimensionality by EPA, Suter, and the 2007 and 2009 National Research Council reports when it evaluated what may be different or similar for risk-assessment endpoints associated with existing versus future biotechnology products.

The term "well understood" in the statement of task was interpreted to mean that the degree of uncertainty in estimates of risk does not preclude a formulation of risk-management options, consistent with the goals and objectives established during the problem-formulation phase of a risk assessment (see EPA, 1998; NRC, 2009; Box 4-1). Although the committee interpreted "well understood" in a risk-analysis context, it noted that "well understood" is a value judgment that in some instances can be informed, at least in part, by the statutory definitions of "safety." The committee also was aware that although a risk analysis for a given product may be "well understood" in one context, it may not be "well understood" in another. As described in this section, the committee examined whether future biotechnology products could pose types of risks different from those associated with existing products (including organisms). It also reviewed risks that are "well understood" and those that may not be in human health and environmental risk assessments.

The Extent to Which Future Products Could Pose Different Types of Risks: Scenarios of Different Use Patterns of Future Biotechnology Products

In 1987, the National Academy of Sciences published a report that stated "[t]he risks associated with the introduction of [recombinant] DNA-engineered organisms are the same in kind as those associated with the introduction of unmodified organisms and organisms modified by other methods" and there is "no evidence that unique hazards exist either in the use of [recombinant] DNA techniques or in the movement of genes between unrelated organisms" (NAS, 1987). What is meant by "same in kind"? The present committee hypothesized that this phrase referred to the final risk-assessment endpoints identified in human health and ecological risk assessments. As these are the kind of assessments that have most commonly been conducted under the auspices of the Coordinated Framework, the committee interpreted its charge to identify "different types of risks" to mean that it should assess the degree to which risk-assessment endpoints identified in human health and ecological risk assessments for existing biotechnology products are likely to be similar to or different from the endpoints that would be selected when assessing risks for future biotechnology

BOX 4-1
Risk-Analysis Refresher

Risk-assessment planning is arguably the most crucial step in the risk-analysis process (EPA, 1998; NRC, 2009). The planning step identifies the goals and objectives of the risk assessment, identifies risk-management options that are under consideration, and identifies the degree of uncertainty that can be tolerated in the risk-management decision. Especially for novel or complex risk assessments, guidance (for example, EPA, 1998) and previous National Research Council reports (NRC, 1996, 2008, 2009) have stressed the significant role public participation can play in supporting effective risk-assessment planning. The subsequent *problem formulation* or scoping phase of an assessment documents the characteristics of the biotechnology product that is the subject of an assessment, its use pattern, the ecosystem or human population potentially at risk, and the endpoints that will be the focus of the assessment. Through one or more risk hypotheses, the *conceptual model* captures the description of the biotechnology product source; what environment it will be used in and how it may move within the environment; and how the product may directly or indirectly interact with specified ecological entities and individuals in specified human populations, as reflected in the risk-assessment endpoints. Key components associated with the temporal and spatial scales to the risk assessment are specified to ensure concordance between exposure and effect analyses to support the risk characterization. In the analysis plan for the risk assessment (the final product in problem formulation), the approaches that will be used to estimate risks qualitatively or quantitatively (for example, through a probabilistic analysis or a combination of exposure and effects) and the anticipated approaches that will be used to evaluate uncertainty in risk estimates are articulated. At the *risk-characterization* phase of the risk assessment, the risk description and risk estimate are provided along with an analysis of the uncertainties associated with the estimate. Assuming the uncertainty in the risk estimate for a specific risk assessment does not preclude informing a specific action to mitigate that risk, the decision-making process concludes, as a practical matter. In such a case, the "risks or lack of risks of a biotechnology product are well understood" in the context of the risk-governance framework and the associated statutory- or voluntary-based definitions of "safety," "reasonable certainty of no harm," or "unreasonable adverse effects," which are the bases of the societal values that inform the selection of endpoints in the assessment.

products. The question in the statement of task about different types of risks was also interpreted by the committee as a request to compare the risk hypotheses linking assumed routes of exposure to possible effects and the spatial and temporal scales used in existing risk assessments with those that may be needed for future products.

In the 2017 update to the Coordinated Framework (EOP, 2017), the regulatory agencies used a number of case studies to illustrate how the updated Coordinated Framework would be applied to products. To a large extent, the case studies review how a particular product would navigate the Coordinated Framework. For all the products reviewed, the route through the Coordinated Framework was relatively well defined. In its evaluation, the committee attempted to articulate scenarios for biotechnology products that could emerge over the next 5–10 years for which the path through the regulatory system would be less clear than for the case-study products. The committee also took into consideration whether risk-assessment endpoints for future biotechnology products would be different from existing biotechnology products.

To organize the scenarios, the committee considered different ways that future biotechnology products might be used or manufactured because exposure to a product or any potential effects on human health or an environment are connected to how a product is used or manufactured. The scenarios include products (including living organisms) that are designed to be released into an open environment and products that are manufactured in contained systems (albeit with limited environmental releases). The committee also examined products with intended or unintended reversible effects as well as products with intended or unintended irreversible effects. It considered exposure of biotechnology products to people or the environment but did not consider human or environmental exposure to compounds produced from biotechnology products.[1] The committee did not attempt to review all available risk assessments and risk-management decisions for existing biotechnology products available in the public domain or prepare problem formulations for possible future biotechnology products. Rather, the committee evaluated several scenarios in which future biotechnology products may be used or manufactured to illustrate key issues and concepts that are required to address the statement of task's question about different types of risk. Some of the scenarios are drawn from two National Academies of Sciences, Engineering, and Medicine reports (NASEM, 2016a,b), while others are drawn from Drinkwater et al. (2014) and presentations made to this committee. The scenarios provided below are intended to be illustrative of the issues that need to be considered to determine similarity in risks of existing biotechnology products with those risks that may be associated with future biotechnology products.

Scenario 1: Contained Products Used in Commercial Manufacturing Facilities That Generate Waste Streams

Future microbial biotechnology products that are used in indoor, contained manufacturing processes and regulated under the Toxic Substances Control Act (TSCA) are likely to be similar in terms of risk-assessment endpoints and the nature of the risk-assessment dimensionality to existing microbial biotechnology products that have already been permitted for manufacturing. Even though such products are intended for use in contained environments, problem formulation would need to include risk-assessment endpoints for the possibility of accidental or intended open release into the environment. For example, if not properly treated, the waste streams from manufacturing processes (for existing and future biotechnology products) may contain engineered biological elements, ranging from genetically engineered organisms to microbial consortia to synthetic DNA. The waste streams themselves may involve either local or interstate activities, depending on how and where the waste streams are treated and distributed.

Under TSCA, EPA has responsibility to address the human health and environmental risks of products released into waste streams. To the extent that control measures in the manufacturing process have a high probability of preventing the release of living organisms into wastewater or solid waste, EPA assumes the risk to humans (that is, adverse health effects) or to the environment (for example, effects on microbial community structure and function) is negligible. EPA also has a list of "pre-approved" microbes that biotechnology-product developers can use; this list is based on risk assessments for these specific microbial species. The human health and ecological risk-assessment endpoints would therefore likely be the same for existing and future products, but risk assessments for a future product—for example, a microbial consortium intended for use in a contained system— may not be as "simple" as current risk assessments for a single, engineered microbe. For example, if there was an accidental release of a living consortium into a waste stream, what is the potential

---

[1]The committee observed that if compounds produced by future biotechnology products were already regulated by EPA (for example, industrial chemicals regulated under the Toxic Substances Control Act) or the U.S. Food and Drug Administration (FDA; for example, cosmetics or food additives under the Food, Drug, and Cosmetic Act that are already sold, distributed, or marketed across state lines) or the compounds (or their proposed uses) are new, existing EPA or FDA processes to assess risks are not different between existing and future compounds. (Of course, there may be unique new chemical compounds created by a biotechnology organism, but the committee concluded the risk analysis of the new chemical compound is outside the report's scope.) However, the committee also observed that the domestic manufacture and use of compounds derived from biotechnology products may not fall under a federal agency's oversight process.

survival and reproduction of the consortium or of each individual microbe? Do the community effects of the consortium affect the ability of individual species within the consortium to survive? How can the consortium be mimicked in laboratory or field studies under a range of environmental conditions that may affect survival and reproduction? Could the consortium or its individual microbes be of concern if consumed by humans? Could the consortium be of concern if consumed by humans even if consumption of its individual species is not a concern? What is the likelihood that the escaped microbes would affect native microbial communities, and is that likelihood different for the consortium versus the individual species? The dimensionality of the risk assessments is thus likely to be more complex than current assessments.

There are also potential regulatory gaps and redundancies. For example, it was not clear to the committee how state regulatory authorities issue national pollutant discharge elimination system permits for potential release of biotechnology products to publicly owned treatment works or to receiving bodies. To what extent would EPA coordinate activities across TSCA, the Clean Water Act, the Resource Conservation and Recovery Act, and the activities of the states (which are, in part, delegated implementation of water and solid waste laws) to minimize redundancy and maximize efficiency in any post-market monitoring? How would potential effects on the environment be monitored, regardless of who has the responsibility to do it? This lack of clarity does not necessarily suggest that there is a risk of concern or that new risks of concern may arise with future biotechnology products; however, the extent to which any adverse effects may be expected by intentional or accidental releases to U.S. waters is not clear.

Scenario 2: Products Manufactured Within Homes for Use by Household Members

Personal or domestic products (such as probiotics, cosmetics, cleaning agents, and antimicrobial pesticides) can be made in traditional factories, but increasingly such products may also be made and used within a home. For example, a purchased kit could allow consumers to engineer organisms to produce a desired chemical, probiotic, or microbe. Combinations of organisms, whether engineered or not, might also be combined in a home environment for use as a nutraceutical or fertilizer.

These types of products may not fall under federal regulatory oversight because presumably they would not be marketed, sold, or distributed and would not cross state boundaries. Risk-assessment dimensionality (time and space), if not risk-assessment endpoints themselves, for future biotechnology products that are manufactured within a home and intended to be used indoors by household members is likely more complex than it is for future products associated with Scenario 1. The nature of exposure pathways and means to estimate environmental releases are less certain as compared to the dimensionality of risk assessments associated with contained manufacturing processes traditionally regulated under TSCA, the Federal Insecticide, Fungicide, and Rodenticide Act (FIFRA), or the Food, Drug, and Cosmetic Act (FDCA). For example, there is likely to be greater variability in geography, use patterns, and disposal in household use than in manufacturing plants and that variability will be difficult for regulators to assess. In addition, the number of children and adults potentially exposed to personal care products or other products used in domestic settings, the number and nature of the settings under which they may be exposed, and the potential adverse effects of the biotechnology products or compounds derived from the products may be less certain as compared to products regulated under TSCA, FIFRA, or FDCA.

As in Scenario 1, the nature and extent of point-source releases of biotechnology products to publicly owned treatment works or non–point source releases to receiving bodies requires a more extensive exposure analysis. Monitoring by the U.S. Geological Survey (USGS) has documented detection of organic wastewater contaminants and pharmaceuticals in public and private drinking-water sources, surface-water receiving bodies, and septic systems (see, for example, Focazio et al.,

2008; Writer et al., 2013; Schaider et al., 2014; Phillips et al., 2015), so it may be reasonable to assume that future biotechnology products manufactured in domestic settings (or the compounds derived from such products) will be released into surface-water or groundwater sources. The results from USGS monitoring studies could provide insights on the dimensionality of the risk hypotheses for future human health and ecological risk assessments. The potential effects of released organisms to the microbial systems in publicly owned treatment works or receiving bodies would be less certain if the potential effects had not been previously characterized. Finally, as with Scenario 1, the dimensionality of risk hypotheses for the disposal scenarios of future biotechnology products will likely be higher and the information to support risk characterization less certain.

Scenario 3: Open-Release, Next-Generation Biotechnology Plants for Agricultural and Other Uses

At the time the committee was writing its report, biotechnology plants consisted of genetically engineered (GE) varieties of a few widely grown crops, such as corn, soybean, and cotton. In total, GE varieties of 10 crop species were grown in the United States in 2015 (NASEM, 2016b). However, the committee anticipated that the number of crop species modified by biotechnology will increase substantially over the next 5–10 years. For example, in the United States, citrus trees engineered to resist citrus greening disease (huanglongbing), which is fatal to the tree, were already in confined trials, and transgenic research was under way to fight Pierce disease in grapes and bacterial spot disease in tomatoes (Ricroch and Hénard-Damave, 2016). Outside the United States, examples of future biotechnology products include GE banana and cassava, for which trials were being conducted in varieties with improved insect and disease resistance and increased nutrient content (Ricroch and Hénard-Damave, 2016). The committee expected that, along with the greater number of GE crop species, the number of engineered genes in a crop would also increase as multiple or more complex traits are targeted and genome-editing techniques such as CRISPR-Cas9 enable certain genetic manipulations to be more readily accomplished. Indeed, a number of crops engineered with genome editing had already been brought to the U.S. Department of Agriculture (USDA) for an "Am I Regulated?" determination (see Table 9-3 in NASEM, 2016b). The committee did not anticipate that the risk-assessment endpoints associated with these new crops would be different from those associated with crops that had already gone through the regulatory process. However, dimensionality, pathways to risk, and the magnitude of risk might change as the synergistic effects of multiple genetic changes could lead to unintended effects in the biochemistry of crops (affecting nutrients, immunogens, phytohormones, or toxicants) or in the phenotypic characteristics of crops due to more complicated epigenetic effects. Off-target effects in genes from base-pair insertions and deletions via genome editing should also be considered.

The committee also expected the use of biotechnology plants to spread beyond agricultural fields. One example would be a novelty product like the glowing plant, discussed in Box 3-4. However, a more important type of plant that the committee thought would be likely to become increasingly common is endangered or locally extinct plants that have been engineered to be able to thrive in natural ecosystems in which they once were widespread. An example discussed in Chapter 2 is the American chestnut, a hardwood tree species native to the U.S. eastern seaboard that has been decimated by a fungal blight introduced in the early 1900s. Genetically engineered resistance to the blight could allow this tree species to grow to maturity and reclaim some of its native range. Ecological risk assessments will be needed for likely inadvertent release of novelty plants into natural ecosystems as well as for the intentional release of plants such as blight-resistant American chestnut.

There may also be regulatory gaps associated with these types of products. For example, if USDA determines that a product is not regulated by virtue of the mechanism used to insert the genetic modification or the source of the genetic material, and that product may be a plant pest or weedy species, there would not be oversight when oversight is warranted.

Scenario 4: Open-Release Microorganisms and Microbial Consortia

Engineered microbial consortia is a potential area of rapid growth in new biotechnology products for open release in the environment for a broad range of markets including mining, bioremediation, and nutrition. As described in Chapter 2, researchers have worked to establish stable synthetic consortia of microorganisms—and the biological principles behind their establishment and maintenance—that could be used as the bases of a wide variety of future applications.

The committee concluded that open release of engineered, naturally occurring or artificial microbial consortia with multiple modifications—some or all of which may be orthogonal—should have a similar suite of risk-assessment endpoints as those used to assess the risks of nonengineered microorganisms, but, as in Scenario 3, the pathways to risk and the magnitude or dimensionality of risk could change. Ecological risk assessments concerning the use of engineered or nonengineered microbial consortia used in bioremediation or biomining would likely address perturbations of native microbial communities including effects on energy flow, horizontal gene transfer, and evolution. The dimensionality of these risk assessments may be more complex with engineered microbes as compared to nonengineered microorganisms, depending on such variables as the use pattern, taxonomic relationships, use of orthogonal genes, environmental conditions within and outside a release site (for example, pH), and use of engineered "kill switches" that terminate the organism when the energy or nutritional sources fall below a certain level. For example, native or artificial consortia designed to alter earthworm digestion of cellulose, change honey bee behavior by manipulating levels of neurotransmitters, or confer nitrogen-fixation properties to nonlegume crops may require a new or expanded suite of risk-assessment endpoints and pathways. If endosymbiotic microorganisms were used to confer nitrogen-fixation properties to nonlegume crops, a human health risk assessment might be needed if the endosymbionts were present in the edible parts of the crop.

Scenario 5: Open-Release Products Designed to Suppress, Eradicate, or Enhance a Target Species Population

Consistent with the National Academies report *Gene Drives on the Horizon: Advancing Science, Navigating Uncertainty, and Aligning Research with Public Values* (NASEM, 2016a) and examples of risk-assessment methods for non-GE biocontrol agents and the release of non-native organisms (Fairbrother et al., 1999; Orr, 2003; Landis, 2004),[2] risk assessments for future biotechnology products that are designed to introduce a new species or suppress or enhance an existing species reflect a high degree of dimensionality and entail a diversity of endpoints at varying levels of biological organization. These assessments will generally require spatially and temporally explicit assessments that address direct and indirect ecological effects and evolutionary effects. Given that some biotechnology products could suppress or enhance a species population at a rate that is faster than natural ecological processes or evolutionary rates, these new products may require the definition of a new suite of risk-assessment pathways.

As an example of the complex pathways that might arise in this type of scenario, a possible risk pathway could be through horizontal gene transfer of the kill-switch mechanism in a gene drive

---

[2]See also the Framework for Assessment described by the U.S. Fish and Wildlife Service learning module on managing invasive plants. Available at https://www.fws.gov/invasives/staffTrainingModule/assessing/introduction.html#part2. Accessed September 13, 2016.

to other species, perhaps to important species in the ecosystem that are beneficial or desired. For example, the possible transfer of a kill switch from a GE organism with a gene drive to that organism's non-GE predator and the detrimental effects of such a transfer would be intermediate risk-assessment endpoints of concern. If the nontargeted predator were to disappear from the ecosystem, its decline could leave a niche open for an invasive species or another pest species. The presence of the harmful species would then be another intermediate endpoint of concern. The results of the harmful species on human health, agriculture, or ecosystems would be ultimate risk-assessment endpoints. Such complex and multilayered risk pathways have been dealt with in other risk analyses for population suppression of mosquitoes (for example, with engineered *Wolbachia*) in fault tree and Bayesian-analysis approaches (Murphy et al., 2010; Murray et al., 2016). Risk assessments for these types of complex pathways may also be able to take advantage from approaches used to assess the risks of non-GE biocontrol agents and the release of non-native organisms (for example, Fairbrother et al., 1999; Orr, 2003; Landis, 2004).[3]

Given the rapid pace of technological change and the ways and environments in which resulting products can be used, it will be important to create and regularly update scenarios such as the ones above to explore emerging risks and the adequacy of the regulatory framework. A recent set of recommendations by the National Academy of Public Administration called for the U.S. government to "systematically integrate foresight into policy development," with an emphasis on the use of scenarios "to consider how different trends and developments may come together in unexpected ways to put policy objectives at risk or create opportunities for more effective action on these objectives" (NAPA, 2016:9,11). Such scenarios should reflect and integrate changes in technologies, capabilities, actors, business models, and risk pathways. Scenarios can be incorporated into a portfolio of approaches, including horizon scanning, to create an early warning system for emerging risks, an approach being explored by the European Union (SEP, 2016).

Risks or Lack of Risks That Are Well Understood

As noted in the introduction of the chapter, the committee interpreted the first question in this part of the statement of task as a request to reflect on the degree to which regulatory human health and ecological risk-assessment endpoints used in risk assessments for existing biotechnology products are likely to be similar to or different from the endpoints that would be selected when assessing risks for future biotechnology products. The second question concerns risks or lack of risks of biotechnology products that are well understood. The term "well understood" was interpreted to mean situations when uncertainty in estimates of risk does not preclude a description of the possible risks consistent with the goals and objectives established during the problem-formulation phase of a risk assessment (see EPA, 1998; NRC, 2009; Box 4-1). However, even with that interpretation, the committee's statement of task was difficult to address given the ambiguity of the phrase "risks or lack of risk." Terms such as "lack of risks," "low risks," "minimal risks," or "acceptable risks" contribute to linguistic ambiguity (NASEM, 2016a) and do not provide a meaningful framework from which to distinguish between scenarios of a product's use in which there are "risks" versus situations where there is a "lack of risks."

Because risk assessments available for existing biotechnology products do not, in general, employ probabilistic estimates of risk, but rather use deterministic expressions of risks (Box 4-2), the committee could not quantitatively address the extent to which risks or the lack of risks are well understood. Existing risk assessments typically characterize risks in a qualitative or deterministic

---

[3]See also the Framework for Assessment described by the U.S. Fish and Wildlife Service learning module on managing invasive plants. Available at https://www.fws.gov/invasives/staffTrainingModule/assessing/introduction.html#part2. Accessed September 13, 2016.

manner, which precludes the means to quantitatively compare risks of existing and future biotechnology products and of biotechnology products to nonbiotechnology products designed for similar use patterns. Advancing quantitative risk assessments will be useful generally, given the characteristics of future open-release products (for example, the mechanisms of action, degree of reversibility and recovery, or movement within ecosystems). There is a need to advance risk-assessment techniques for potential adverse outcomes that have not been rigorously addressed for both biotechnology products and environmental stressors[4] in general (NRC, 2013; NASEM, 2016a). Given the nature of the use patterns for future open-release products, spatially and temporally explicit risk assessments will also facilitate a more insightful identification of risk patterns.

Risks That Are Well Understood

Although it was not possible to quantitatively determine risks that are well understood, a future biotechnology product that is based on a similar genetic modification and has a similar use pattern as an existing biotechnology product with a safe-use record likely has a risk profile similar to that of the existing product.

The scenarios in the above section could be used for pilot projects to develop probabilistic estimates of risks for existing biotechnology products and thereby provide the means to compare the likelihood of adverse effects from future biotechnology products to the likelihood of adverse effects from existing biotechnology and nonbiotechnology products, assuming risk assessments for future products incorporate probabilistic methods. Such analyses would help identify high-priority information needs to reduce uncertainty in risk estimates and inform the classification of comparable products based on the nature of risk-assessment endpoints, dimensionality of risk assessments, and the probabilities of adverse effects. Pilots would be particularly helpful for products intended for wide-area environmental release in low-management environments (for example, open-release organisms with gene drives or genetically engineered bacteria for bioremediation or fuel production). Estimated probabilities of immediate, medium, and long-term environmental and human health risks would be appropriate (Suter, 2007; Warren-Hicks and Hart, 2010; NRC, 2013; NASEM, 2016a).

For new biotechnology products without comparators, risk estimates could be lower than, similar to, or higher than such estimates for existing biotechnology or nonbiotechnology products based on the design of the future products and their use patterns.

Risks That Are Not Well Understood

Although risk-assessment endpoints for human health and environmental effects for existing and future biotechnology products will likely be similar (assuming, for example, similar manufacturing controls, use patterns, and properties of the products), the magnitude of the risks may change, the pathways could be more complex and multidimensional, and existing risk assessments have limitations. It is not always clear in existing risk assessments how assessment endpoints were selected or how the dimensionalities of the risk assessments were considered. Furthermore, existing risk assessments are generally comparative risk assessments—that is, they rely in large part on the comparison of a new biotechnology product to an existing nonbiotechnology product. Future products of biotechnology will involve more complex comparisons that, for example, potentially encompass multigene traits in consortia deployed in novel environments and management scenarios, and the new products may not have a counterpart or precedent to allow a ready means for an "as safe as" comparison to a product that already exists. The comparative risk-assessment

---

[4]A *stressor* is any agent or actor with the potential to alter a component of the ecosystem (NASEM, 2016a).

BOX 4-2
Deterministic Versus Probabilistic Risk Assessment

Typically deterministic risk estimates are used in human health and ecological risk assessments for biotechnology products. Deterministic risk assessments provide a single solution for exposure and effects variables (for example, a comparison of a single point estimate of exposure for a selected magnitude of effect). Often these assessments provide a risk description that is based on the perspective that assumptions used in exposure and effect analyses will provide an adequate margin of safety. The deterministic approach provides incomplete characterization of the variability and epistemic uncertainty (lack of or incomplete knowledge) in risk estimates. Imprecise language in deterministic risk estimates and risk characterizations also contributes to linguistic ambiguity in risk assessments.

The lack of a probabilistic approach for estimating risk and evaluating variability and uncertainty in risk estimates can undercut the overall regulatory process (as summarized by Sunstein, 2005; Warren-Hicks and Hart, 2010). Lack of clarity on the probability of an adverse outcome contributes to disagreements between regulatory authorities, the regulated community, and stakeholders and other interested publics as to whether or not assumptions in a deterministic assessment are sufficient to meet a safety standard; the nature and extent of additional data needed to address uncertainties that are acknowledged in a deterministic assessment; and the nature and extent to which safety or uncertainty factors influence a risk estimate.

Risk analyses that make use of stochastic methods describe the range of possible outcomes and the dependence of outcomes on estimated parameters or assumptions. These analyses can increase transparency about how much is known about the risks of a product and where more data and information may be needed. Several previous National Academies committees have recommended increased use of probabilistic (quantitative) risk analyses (NRC, 2009, 2013; NASEM, 2016a). A National Research Council (2013) report noted that a deterministic risk estimate provides no information on whether the error in the risk estimate is consistent with the needs of risk managers. For these reasons, that report also concluded that it is not possible to ascertain the significance of potential differences in adverse effects occurring when comparing deterministic risk estimates for different scenarios (for example, different products or different species). Using an example from that 2013 report, expressing risk of a product's use as a 20-percent ± 10-percent probability of a 25-percent reduction in the population growth rate of a specified assessment species in a specified location provides the means to compare risks for different use patterns of the same product or to compare risks across different products. In this way, probabilistic assessments provide the means to quantitatively evaluate variability and the effects of epistemic uncertainties and can support risk descriptions that minimize linguistic ambiguity.

This is not to say that probabilistic risk assessments are warranted or possible for all circumstances. There may be instances in which (a) a deterministic risk assessment based on well-characterized, worse-

---

process may, however, be more easily applied to human health risk assessments than to ecological risk assessments. Finally, it is difficult and not typical to incorporate social and cultural factors into existing risk assessments, yet these factors may ultimately affect human health and ecological risk (for example, social systems and variability in use of products). In addition, social and cultural values, including social and cultural risks, are not usually included in regulatory risk assessments and are difficult to assess.

Products Without Comparators. Existing biotechnology risk assessments are guided by comparative risk approaches and are informed by comparisons to nonbiotechnology counterparts that help establish the "as safe as" criteria used by regulatory decision makers. As described in Chapter 2, existing biotechnology products have typically involved a host (the organism into which new material is introduced) and a source (the organism from which the new material was taken). For example, for a GE insect-resistant variety of corn, the corn is the host organism and the bacterium *Bacillus thuringiensis* (*Bt*) is the source organism. In the comparative risk-assessment paradigm, the GE corn variety is compared against a genetically similar variety of corn that does not contain the GE

case assumptions supports the safety standard associated with a risk-management decision criteria, (b) field studies or monitoring studies clearly demonstrate observed adverse outcomes are related to a specific stressor, or (c) the costs of implementing risk-mitigation steps to reduce risks while maintaining benefits are low and less than the costs needed to refine an assessment (Sunstein, 2005; Warren-Hicks and Hart, 2010). Furthermore, the Institute of Medicine recommended that in-depth uncertainty analyses may not be warranted if perfect information would not change the decision; that is, "[t]he effort to analyze specific uncertainties through probabilistic risk assessment or quantitative uncertainty analysis should be guided by the ability of those analyses to affect the environmental decision" (IOM, 2013:15). There may also be instances in which there are insufficient data or knowledge to implement equations in risk models or reasonably inform probability distribution functions or the relationships between such functions (Warren-Hicks and Hart, 2010). There are also examples in the literature describing how components of existing ecological risk assessments could be enhanced to provide a probabilistic risk estimate (Wolt and Peterson, 2010). In this regard, the 2009 National Research Council report noted that there is a continuum of approaches for characterizing uncertainty (for example, use of default assumptions; qualitative description of uncertainty; bounding values, interval analysis, and sensitivity analysis; and probabilistic analyses) that can be used consistent with goals established in problem formulation (NRC, 2009). The 2013 Institute of Medicine report also noted there can be situations that present deep uncertainty; this may occur when underlying environmental processes are not understood and there is fundamental disagreement among scientists that is not likely to be reduced by additional research within the time period for which a regulatory decision must be made.

If the potential adverse effects are reversible and not catastrophic, but there is some low probability the adverse effect could occur based on a deterministic assessment with high margins of safety for a specific use pattern (for example, a limited field study with confinement and containment), it may be reasonable to allow the generation of needed data to inform a subsequent probabilistic assessment (Sunstein, 2005). This phased approach to generating data to support increasingly refined quantitative risk assessments was articulated in the National Academies report on gene drives (NASEM, 2016a).

Selection of the appropriate risk-assessment approach (deterministic or probabilistic) can help address the problem of disproportional efforts in risk management; that is, it can help prevent the problem of second-order risk—the risk of missing a significant risk versus the risk of overanalyzing a negligible risk. However, increasing use of probabilistic risk assessments will require careful attention to risk communication to ensure risk managers and interested and affected parties understand uncertainties when formulating risk-management options; the 2013 Institute of Medicine report is an example review that addresses formulating regulatory decisions under uncertainty.

trait. The two varieties are compared in terms of nutrient and chemical composition to ascertain if there are unintended differences (the GE trait is an intended difference). This comparison includes toxicity testing and allergenicity screening to help identify any potential human health issues related to intended or unintended effects from the GE trait.

With the advancement of biotechnology from recombinant-DNA technology to genome engineering, the use of comparators is becoming more challenging, even for GE crops (EFSA, 2011). Furthermore, transformations can be made in host organisms that are less well characterized than corn. For example, genome-editing technologies allow product developers to make changes in genomes of nearly any host organism for which there is a genome sequence available, from microbes to insects to mammals (Reardon, 2016). There may not be baseline data on the nontransformed counterpart host. Furthermore, novel gene sequences—including synthetic ones—can be introduced into host organisms; there may be no nonbiotechnology product to which they can be compared.

If potential off-target effects of new technologies, such as genome editing, are similar to those that occur naturally—for example, point mutations or epigenetic changes—and the probability of off-target effects is not substantially different from the background rate of such changes, then any *additional* risk (that is, beyond that associated with the intended target of genome-editing technology) is low. On the other hand, the range of genome-editing techniques is rapidly expanding, and if certain new technologies cause off-target effects that do not typically occur naturally or increase the number of such changes, then this may pose an additional challenge for comparative risk assessment. Note that conventional-breeding techniques such as mutation breeding in plants could increase the extent and the range of types of off-target effects to a much greater extent than existing genome-editing techniques (NASEM, 2016b). With rapid changes in and lowering costs of genome sequencing and genome-interrogating technologies (see Chapter 2), genome-wide identification of off-target effects is increasingly straightforward. However, what remains a challenge is evaluating whether any off-target effects pose a risk because of the difficulty in obtaining proper comparators. Finally, for organisms made by inserting entirely new pathways of genes derived from multiple unrelated sources or consortia of organisms engineered with multiple genes, how a product developer or regulator would conduct protein toxicity testing is complicated. Synergistic effects of multiple genes and organisms compound such testing, and it might not be accurate to test each organism or gene separately. In these cases, basic biology and ecology studies may be necessary to develop baseline environmental behaviors, as in new knowledge about gene transmissibility, persistence in the environment, and toxin production before small-scale environmental release.

On a related note, products without comparators do not have a clear path already charted through the U.S. regulatory system. The degree to which risks associated with a new type of product are "well understood" is a subjective determination, but formal uncertainty analyses can help determine when a reduced level of regulatory scrutiny may be warranted. Use of a biological component or system that has a history of safe use provides an existing regulatory path with known nonbiotechnology comparators and clear risk-assessment endpoints, whereas using an unfamiliar component or system does not. Future biotechnology products may be unique and therefore lack adequate precedents. For example, the GE mosquito developed by the company Oxitec was a first-of-its-kind product; it was genetically engineered to carry a gene that would render its offspring sterile with the goal of effectively eradicating the mosquito population (*Aedes aegypti*) that, among other things, carries dengue, chikungunya, and Zika viruses. The developer and the regulatory agencies both shared uncertainty on the front end as to whether the GE mosquito should be regulated by EPA under FIFRA as an insecticide (because it kills mosquitoes) or by the U.S. Food and Drug Administration (FDA) as a new animal drug (the pathway that was used for genetically engineered salmon). After a number of years of agency deliberations, in August 2016 FDA ultimately released a final environmental assessment and a finding of no significant impact, agreeing with Oxitec's environmental assessment that its proposed field trial would not have significant impacts on the environment. However, FDA's assessment was not a quantitative, probabilistic one and was not based on field data looking for harm to nontarget species or to ecosystems; field data from other countries only looked at the efficacy of population suppression (FDA, 2016). The committee heard a similar concern about a lack of clarity with regards to the regulatory path for engineered microbes. The microbes were designed for open-environmental release to extract gold and copper from low-grade ore. An invited speaker emphasized that there was ambiguity about which regulatory agency would be responsible for the product, what data would be required by the agency, and what nontransformed host would serve as the comparator in a risk assessment (DaCunha, 2016).

Social and Cultural Effects. Methods to incorporate social and cultural values into risk analysis are limited because they often cannot be put on the same scale as health risks, environmental externalities, and monetized costs and benefits. However, some risk-analysis frameworks have been developed to incorporate values of various publics during the problem-formulation or risk-management stage, when other options are compared to the proposed action of releasing a techno-

---

BOX 4-3
Public Confidence

Important quantitative correlations have been found between public confidence in oversight of bio-technology products and opportunities for public input, incentives for regulatory compliance, and the strength of data requirements for regulation (Kuzma et al., 2009a,b, 2010). These studies highlight the importance of reducing complexity and uncertainty for minimizing financial burdens on small product developers; consolidating multiagency jurisdictions to avoid gaps and redundancies in safety reviews; consumer benefits for advancing acceptance of products; rigorous and independent premarket and post-market assessment for environmental safety; early public input and transparency for ensuring public confidence; and the positive role of public input in system development, informed consent, capacity, compliance, incentives, and identifying data requirements (Kuzma et al., 2009a,b, 2010). An earlier National Research Council report looking at GE insect-resistant crops also identified gaps and redundan-cies, lack of capacity, low requirements for field data in the case of USDA regulation, lack of transparency due to confidential business information, and little to no post-market monitoring outside of EPA's FIFRA regulation (NRC, 2000).

Early engagement in public discussions regarding novel, disruptive, and controversial technologies is important for (a) ethical or normative reasons, such as procedural justice and informed consent (for example, having a voice and choice especially in a democracy); (b) increasing the legitimacy of policy processes (through improved transparency, integrity, and credibility); (c) increasing the knowledge base on which decisions are made, which can often lead to better decision outcomes because people have knowledge about systems into which biotechnologies are deployed that technological developers and regulators do not (NRC, 1996; NASEM, 2016a); and (d) capacity building and learning, in that public engagement helps to increase future capacities for participants to work within science and technology policy processes that develop an enhanced sense of civic ownership, civic commitment, and civic aware-ness (Selin et al., 2016).

---

logical product (Nelson et al., 2004). Previous National Academies reports have also emphasized the importance of public participation in the risk-analysis process, particularly in risk character-ization (which involves complex, value-laden judgments) and problem formulation, where public involvement can improve acceptance of the analysis (Box 4-3) and improve the analysis for the purposes of risk management (NRC, 1996, 2008, 2009; NASEM, 2016a).

Future products of biotechnology will be more complex in terms of their interactions with their environment and society, and more research may be needed to develop methods for governance systems that integrate ethical, cultural, and social implications into formulation of risk-assessment endpoints and risk characterization in ways that are meaningful. At the same time, it may not be feasible or even justified for all new biotechnology products—for instance, products with which there is already familiarity or products that will not be released into the environment. Genetically engineered organisms used in the research laboratory to develop new chemical synthesis methods are not likely to require the same level of public dialogue according to decision criteria for public engagement proposed by several scholars and think-tanks as will products that have more uncer-tainty associated with them, such as organisms with gene drives (Funtowicz and Ravetz, 1995; NRC, 1996; Renn, 2005; NASEM, 2016a).

EXISTING FEDERAL CAPABILITIES, EXPERTISE, AND CAPACITY

To address risks that are not "well understood" and to oversee the profusion of products antici-pated from the discussion in Chapter 2, federal agencies need to be prepared with the appropriate scientific capabilities, expertise, and capacity to conduct regulatory science. On the basis of definitions provided by FDA[5] and the Society for Risk Analysis, the committee understood *regulatory science* to involve developing and implementing risk-analysis methods and maximizing the utility of risk analyses[6] to inform regulatory decisions for biotechnology products, consistent with human health and environmental risk–benefit standards provided in relevant statutes. Regulatory science includes establishment of information and data quality standards, study guidelines, and generation of data and information to support risk analyses. It can also include the development of risk-mitigation measures and the development and implementation of safety training and certification programs to help ensure the intended benefits of products are realized and risks to workers, users, and the environment are minimized. Individuals in government, industry, academia, and nongovernmental organizations that contribute to the advancement of regulatory science have degrees across disciplines in the natural, socioeconomic, and computational sciences, engineering, and public policy.

Federal capacity is not limited to the agencies that participate in the Coordinated Framework. To assess the capacity of the federal government to regulate future biotechnology products, the committee looked at the existing capabilities in the workforce, the available external resources that could be drawn upon by the agencies, and the present investment in key tools for biotechnology-product evaluation. The committee also noted current opportunities to enhance capability and capacity through interactions across the federal agencies and through partnerships with developers, nongovernmental organizations, and academia.

Existing Scientific Capabilities in the Federal Workforce

The committee made use of the FedScope database within the U.S. Office of Personnel Management to ascertain the types of expertise within the regulatory agencies. The database provided information about the trends in number of staff with each type of expertise for the fiscal years 2011–2015.[7] From the list of professional occupations provided in the database, the committee selected 33 that it surmised would be necessary to the regulatory responsibilities of EPA, FDA, or USDA–Animal and Plant Health Inspection Service (APHIS) personnel; the search returned results for at least one agency in 25 of the occupations.[8] Figures 4-1 through 4-6 show the employment trends for the occupations in each agency. There are two figures per agency to address issues of scale. Professions or years which returned a value of "NA" in the database are not included in the figures.

The committee recognizes that the data provided in Figures 4-1 through 4-6 are not as specific as would be desired. The committee was primarily interested in the expertise available with FDA's Center for Food Safety and Applied Nutrition and Center for Veterinary Medicine, EPA's Office

---

[5]U.S. Food and Drug Administration. Advancing Regulatory Science: Moving Regulatory Science into the 21st Century. Available at http://www.fda.gov/ScienceResearch/SpecialTopics/RegulatoryScience/default.htm?utm_campaign=Goo. Accessed December 13, 2016.

[6]According to the Society for Risk Analysis, *risk analysis* defined broadly includes "risk assessment, risk characterization, risk communication, risk management, and policy relating to risk, in the context of risks of concern to individuals, to public and private sector organizations, and to society at a local, regional, national, or global level" (SRA, 2015).

[7]FedScope employment trends are available at https://www.fedscope.opm.gov/etrend.asp. Accessed December 12, 2016.

[8]The professions listed in the FedScope database correspond to the qualification standards for federal positions used by the Office of Personnel Management. The committee searched the following professions: general natural resources management and biological sciences, microbiology, pharmacology, ecology, zoology, physiology, entomology, toxicology, botany, plant pathology, plant physiology, horticulture, genetics, soil conservation, forestry, soil science, agronomy, fish and wildlife administration, fish biology, wildlife biology, animal science, environmental engineering, bioengineering and biomedical engineering, agricultural engineering, chemical engineering, chemistry, forest products technology, food technology statistics, social science, sociology, economist, and veterinary medical science. No results were returned for the professions plant physiology, horticulture, genetics, soil conservation, forestry, agronomy, sociology, and fish and wildlife administration.

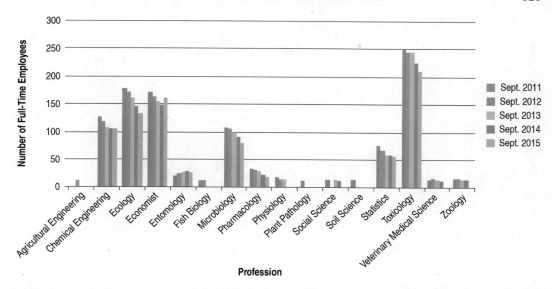

FIGURE 4-1 Professions of interest at the U.S. Environmental Protection Agency with fewer than 300 employees, fiscal years 2011–2015.
DATA SOURCE: U.S. Office of Personnel Management. FedScope database, employment trend cubes. Available at https://www.fedscope.opm.gov/etrend.asp. Accessed December 12, 2016.

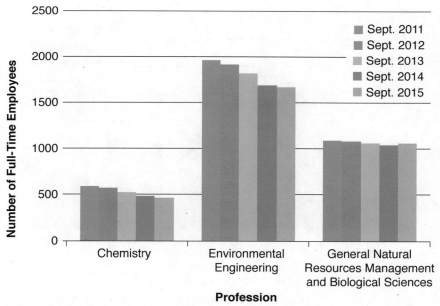

FIGURE 4-2 Professions of interest at the U.S. Environmental Protection Agency with more than 400 employees, fiscal years 2011–2015.
NOTE: *General Natural Resources Management and Biological Sciences* includes employees hired with expertise in biological sciences, agriculture, natural resource management, chemistry, or related disciplines appropriate to the position.
DATA SOURCE: U.S. Office of Personnel Management. FedScope database, employment trend cubes. Available at https://www.fedscope.opm.gov/etrend.asp. Accessed December 12, 2016.

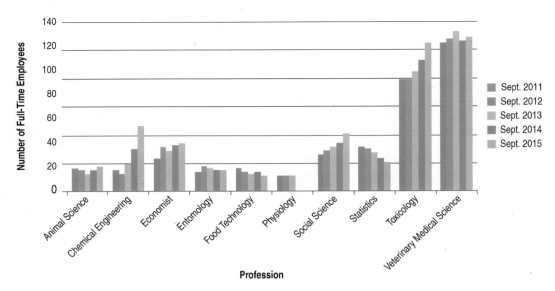

FIGURE 4-3 Professions of interest at the U.S. Food and Drug Administration with fewer than 150 employees, fiscal years 2011–2015.

DATA SOURCE: U.S. Office of Personnel Management. FedScope database, employment trend cubes. Available at https://www.fedscope.opm.gov/etrend.asp. Accessed December 12, 2016.

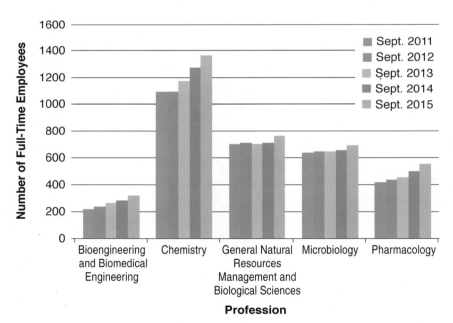

FIGURE 4-4 Professions of interest at the U.S. Food and Drug Administration with more than 200 employees, fiscal years 2011–2015.

NOTE: *General Natural Resources Management and Biological Sciences* includes employees hired with expertise in biological sciences, agriculture, natural resource management, chemistry, or related disciplines appropriate to the position.

DATA SOURCE: U.S. Office of Personnel Management. FedScope database, employment trend cubes. Available at https://www.fedscope.opm.gov/etrend.asp. Accessed December 12, 2016.

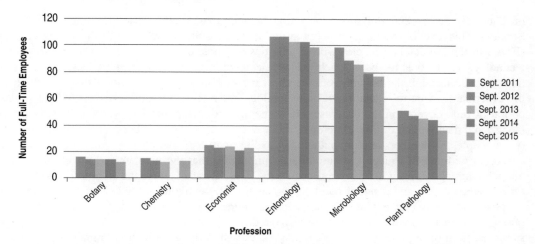

FIGURE 4-5 Professions of interest at the U.S. Department of Agriculture's Animal and Plant Health Inspection Service with fewer than 150 employees, fiscal years 2011–2015.

NOTE: This figure has been updated since the prepublication release. The number of full-time employees in the prepublication version of the report was incorrect.

DATA SOURCE: U.S. Office of Personnel Management. FedScope database, employment trend cubes. Available at https://www.fedscope.opm.gov/etrend.asp. Accessed December 12, 2016.

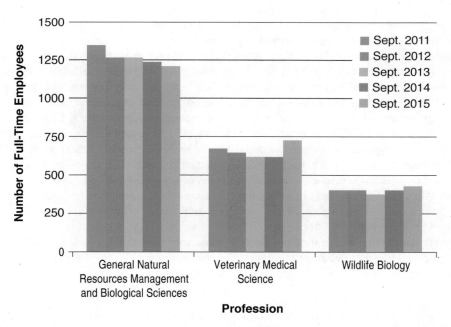

FIGURE 4-6 Professions of interest at the U.S. Department of Agriculture's Animal and Plant Health Inspection Service with more than 300 employees, fiscal years 2011–2015.

NOTE: *General Natural Resources Management and Biological Sciences* includes employees hired with expertise in biological sciences, agriculture, natural resource management, chemistry, or related disciplines appropriate to the position.

DATA SOURCE: U.S. Office of Personnel Management. FedScope database, employment trend cubes. Available at https://www.fedscope.opm.gov/etrend.asp. Accessed December 12, 2016.

of Chemical Safety and Pollution Prevention, and USDA–APHIS's Biotechnology Regulatory Services. This level of resolution was not available; as a result, the number of full-time employees (FTEs) in the regulatory agencies whose activities are most directly relevant to nonmedical biotechnology products is likely well below those depicted in the figures. Nevertheless, the information returned through the committee's search of the FedScope database shows that, for much of the expertise relevant to regulation of biotechnology, the number of employees has fallen over the period of time covered in the database. For some applied disciplines important to understanding the use and deployment of biotechnology products, the decrease in numbers is a cause for concern (for example, the numbers of plant pathologists, soil scientists, fisheries biologists, and agricultural engineers within EPA) given the committee's prediction that the number and novelty of future biotechnology products will increase. Because the FTE data available to the committee likely overestimate the breadth and depth of regulatory agency staff, the trends in the federal workforce raise the possibility there may not be sufficient capacity to meet future regulatory demands. There is the possibility of using contract scientists to manage the day-to-day workload of review, analysis, and summarization of regulatory submissions. Agencies within the Coordinated Framework have and do maintain contract personnel and have this built into agency budgets. Proper utilization of such outside expertise will require appropriate training and oversight to ensure the quality of regulatory reviews is maintained.

Enhancing Federal Expertise Through External Scientific Advisory Committees

The agencies can and have supplemented their internal expertise by making use of external advisory committees. Independent federal advisory committees can not only assist with scientific peer review on issues of human health and ecological risk assessment but can also provide input on issues such as risk-assessment methodology, environmental modeling, life-cycle analyses, sustainability, and environmental justice, among others, which may become more relevant in assessing risks and benefits of future products if the conception of risk is broadened (see Box 3-1). EPA has used such committees to provide independent, external peer review, with public comment, on specific, well-defined risk-assessment methodology and assessment issues for specific biotechnology products.[9] In the context of pesticidal biotechnology products, EPA has used its science advisory panels (SAPs) to review risk assessments and resistance management plans for *Bt* plant-incorporated protectants since the mid-2000s.[10] In 2014, the FIFRA SAP reviewed EPA's proposed problem-formulation approach for RNA interference (RNAi) technology in anticipation of future human health and ecological risk assessments of specific RNAi products (EPA, 2014a). Consistent with an iterative process in advancing risk-assessment approaches for a new technology, EPA convened a SAP in September 2016 to review a draft assessment for a specific RNAi plant-incorporated protectant (EPA, 2016c).

Other agencies acting within the Coordinated Framework have utilized programmatic environmental impact statements to elicit public input as a means to consider need for future regulation for biotechnology products. For instance, USDA–APHIS undertook programmatic environmental

---

[9]See, for example, EPA FIFRA Scientific Advisory Panel for pesticides, including biotechnology products (available at https://www.epa.gov/sap, accessed January 14, 2017); the EPA Science Advisory Committee on Chemicals, including biotechnology products regulated under TSCA (TSCA Scientific Peer Review Committees, available at https://www.epa.gov/tsca-peer-review, accessed January 14, 2017); and USDA's Advisory Committee on Biotechnology & 21st Century Agriculture (available at http://www.usda.gov/wps/portal/usda/usdahome?contentidonly=true&contentid=AC21Main.xml, accessed January 14, 2017).

[10]See Scientific Advisory Panel Meetings on Issues Related to PIPs (External Peer Review). Available at https://www.epa.gov/regulation-biotechnology-under-tsca-and-fifra/overview-plant-incorporated-protectants#scientific. Accessed January 14, 2017.

impact statement activities in 2004 for the purpose of rulemaking to determine if their regulatory remit for biotechnology regulation could be clarified and expanded (USDA–APHIS, 2004), and it was involved in a similar process at the time the committee was writing its report (USDA–APHIS, 2016). In addition, USDA–APHIS has sponsored third-party activities to gain expert input that was subsequently reflected in views and approaches used in biotechnology risk assessment (Traynor and Westwood, 1999). USDA–APHIS has also directly sponsored expert workshops (Rose et al., 2006). There also have been expert meetings hosted by the former interagency Agricultural Biotechnology Risk Analysis Task Group to understand research priorities (NSTC, 2007). Agencies responsible for regulating future biotechnology products can also build from experience to proactively gain advice and input on proposed risk-assessment approaches for other future products and assessment techniques in areas like nanotechnology (EPA, 2009), computational toxicology (EPA, 2011), spatially explicit ecological assessments (EPA, 2015), and pollinator protection (EPA, 2012).

EPA, FDA, and USDA–APHIS have used independent external scientific input and review several times in the past. For example, EPA sought input from the National Research Council (NRC, 2012) to provide advice on future environmental science and engineering challenges and technological advances and to assess the overall capabilities of the agency to meet emerging and future mission challenges. One of the recommendations was to "[e]ngage in a deliberate and systematic 'scanning' capability involving staff from [EPA's] ORD [Office of Research and Development], other program offices, and the [EPA regional offices]. Such a dedicated and sustained 'futures network' (as EPA has called groups in the past with a similar function), with time and modest resources, would be able to interact with other federal agencies, academe, and industry to identify emerging issues and bring the newest scientific approaches into EPA" (NRC, 2012:11). Consistent with this horizon-scanning recommendation, in 2016 EPA shared a preliminary view of emerging and potential issues, which included the inevitability of transformational biotechnology products (Greenblott et al., 2016).

The benefit of this input from external experts is reflected in guidance that lays the groundwork for implementation of strengthened approaches for risk assessment. For instance, guidance for identifying and selecting ecological risk-assessment endpoints (including those that address ecological goods and services) across biological, spatial, and temporal scales have been developed (for example, EPA, 2003, 2016g), and EPA has summarized experiences gained in several case studies that assessed the human health and ecological risks of exposure to combinations of disparate chemical, biological, and physical stressors. These case studies also highlight the importance of engaging stakeholders throughout the risk-assessment phases and risk management (Gallagher et al., 2015; see also Box 4-3). The U.S. Fish and Wildlife Service (FWS), the National Marine Fisheries Service (NMFS), USDA, and EPA were in the process of implementing recommendations from the 2013 National Research Council report *Assessing Risks to Endangered and Threatened Species from Pesticides* at the time the committee was writing its report.[11] EPA's Risk Assessment Forum has developed peer-reviewed white papers on the use of probabilistic human health and ecological risk assessments (EPA, 2014b) in response to the 2013 Institute of Medicine report on uncertainty in environmental decision making (IOM, 2013). Probabilistic human health risk assessments are performed on a routine basis for food-use pesticides (EPA, 2016a) based on advice from the FIFRA SAP. Probabilistic methods have been developed for some ecological risk-assessment scenarios (NRC, 2013; EPA, 2016e) and have been employed in a limited number of cases. Ongoing research to advance computational toxicology techniques, including high-throughput screening, illustrates EPA's commitment to advance 21st-century approaches to assess chemical stressors (NRC, 2007)

---

[11]Implementing NAS Report Recommendations on Ecological Risk Assessment for Endangered and Threatened Species. Available at https://www.epa.gov/endangered-species/implementing-nas-report-recommendations-ecological-risk-assessment-endangered-and. Accessed September 13, 2016.

and develop and employ the information-technology infrastructure needed to manage and analyze large data sets (for example, see EPA, 2016f).

EPA, USDA, and USGS already maintain and employ large geospatial data sets to support human health and ecological risk assessments[12] and agriculture and natural resource research and management within the governance of the Federal Geographical Data Committee (NRC, 2013). Research is ongoing to develop the data, models, and tools to expand community stakeholders' capabilities to consider the social, economic, and environmental impacts of decision alternatives on community well-being; develop the causal relationships between human well-being and environmental conditions; develop and implement monitoring designs and indicators to support national, regional, and state reports of environmental condition; and advance tools and metrics to support life-cycle analyses (Yeardley et al., 2011; EPA, 2016d). The agencies also have established processes to address cross-cutting research and scientific issues, including dialogue with stakeholders,[13] which could be expanded to complement cross-cutting issues relevant to the Coordinated Framework and future research and risk-assessment needs.

Another example of an external advisory group is the EPA Pesticide Program Dialogue Committee (PPDC),[14] which was established in 1995 as a forum for a diverse group of stakeholders—environmental and public-interest groups, pesticide manufacturers, trade associations, commodity groups, public health and academic institutions, federal (including USDA, FDA, FWS, NMFS, and the Centers for Disease Control and Prevention) and state agencies, and the general public—to provide feedback to EPA on various pesticide regulatory, policy, and program implementation issues. The PPDC has provided advice to EPA in implementing far-reaching changes in risk-assessment and risk-management approaches mandated with passage of the Food Quality Protection Act in 1996 and provided input on issues including implementation of 21st-century toxicology testing and nonanimal testing alternatives, endangered species and pollinator protection options, classification systems for reduced-risk pesticides, and approaches for documenting label claims, among others. The PPDC also provided feedback on EPA's development and implementation of public review and comment processes for proposed new pesticide registration decisions (EPA, 2016b) and the reevaluation of registered pesticides (including problem formulation, draft risk assessments, and proposed regulatory decision steps[15]). This EPA advisory committee could be employed to help guide the development of a governance approach for pesticidal biotechnology products in conjunction with the Coordinated Framework.

Independent external advice and input can help expand the ability of a federal agency to meet its future scientific challenges. Successful implementation of recommendations from an advisory committee is contingent on the breadth and depth of an agency's workforce. Downward trends in the staffing of certain areas of expertise at regulatory agencies (summarized in Figures 4-1 through 4-6) raise concerns that the staff may not have sufficient time or skills to take advantage of external advice and prepare for the future.

---

[12]EnviroAtlas. Available at https://www.epa.gov/enviroatlas. Accessed October 11, 2016.

[13]See, for example, Computational Toxicology Communities of Practice, available at https://www.epa.gov/chemical-research/computational-toxicology-communities-practice, and Pesticide Environmental Modeling Public Meeting–Information, available at https://www.epa.gov/pesticide-science-and-assessing-pesticide-risks/environmental-modeling-public-meeting-information. Both accessed January 10, 2017.

[14]Pesticide Program Dialogue Committee. Available at https://www.epa.gov/pesticide-advisory-committees-and-regulatory-partners/pesticide-program-dialogue-committee-ppdc. Accessed January 14, 2017.

[15]Registration Review Process. Available at https://www.epa.gov/pesticide-reevaluation/registration-review-process. Accessed January 14, 2017.

Federal Research Funding to Advance Risk Analysis on Future Biotechnology Products

In an attempt to better ascertain the nature and extent of federal research designed to support risk analyses of biotechnology products, the committee solicited input from relevant agencies through a request for information (RFI) keyed to programmatic work addressing risk analysis for products of biotechnology. The questions posed in the RFI (see Appendix C) were derived, in part, from the report *Creating a Research Agenda for Ecological Implications of Synthetic Biology*, published in 2014 following two workshops organized by the Massachusetts Institute of Technology Program on Emerging Technologies and the Woodrow Wilson Center's Synthetic Biology Project (Drinkwater et al., 2014) and from a workshop and Delphi study on synthetic-biology governance funded by the Sloan Foundation and hosted by North Carolina State University's Genetic Engineering and Society Center (Roberts et al., 2015). The committee was interested in programmatic work related to fundamental and applied research efforts that can inform human, animal, and ecological risk assessments and socioeconomic costs and benefits. Research related to potential risks of future human drugs or medical devices was not included in the committee's statement of task and therefore was not part of this RFI, except to the extent such research may be broadly applicable to other biotechnology products.

The committee sent the RFI to 28 federal offices and received responses from 17. Twelve of the 17 had information to share that was relevant to the committee's request. The RFI recipients are listed in Appendix C. The RFI specifically asked about the ongoing research the agencies had with regards to

- The nature and extent to which future biotechnology products were similar to or different from nontransformed (nonbiotechnology) products serving as comparators.
- Off-target gene effects[16] and phenotype characterization of future biotechnology products.
- Impacts of future biotechnology products on nontarget organisms.
- Gene fitness, gene stability, and propensity for horizontal gene transfer in future biotechnology products.
- Measures designed to control organismal traits and mitigate risk in the event of intentional or accidental release of future biotechnology products.
- Life-cycle analysis of future biotechnology products.
- Monitoring and surveillance of future biotechnology products.
- Modeling to inform risk-based hypotheses, collect data to reduce uncertainties, and provide findings or predictions in risk characterization with regards to future biotechnology products.
- Economic costs and benefits of future biotechnology products.
- Social costs and benefits of future biotechnology products.

The committee also included an "other" category to catch any other areas of research not described above in the event any of the agencies receiving the RFI had additional information to share. Information falling into this category is not included in the following figures but is described within the text.

EPA reported it had no ongoing research directly tied to any one area of the RFI. However, the agency did provide the committee with information regarding efforts within its Office of Research

---

[16]It is important to distinguish between nontarget effects and off-target effects. *Nontarget effects* are unintended, short- or long-term consequences for one or more organisms other than the organism intended to be affected by an action or intervention. Concern about nontarget effects typically centers around unforeseen harms to other species or environments, but nontarget effects can also be neutral or beneficial. *Off-target effects* are unintended, short- or long-term consequences of an intervention on the genome of the organism in which the intended effect was incorporated. See also NASEM (2016a).

and Development that would enhance risk-assessment capabilities and could be applied to biotechnology. It is worth noting that EPA previously funded intramural and extramural biotechnology research programs, but these activities were discontinued in 2012. At the time the committee was writing its report, EPA possessed some capacity and capability for implementing probabilistic ecological risk assessments (as noted above) but had done so only on a limited basis for nonbiotechnology pesticides; models appropriate for ecological risk assessments of biotechnology products had not been developed.

USDA reported a total of approximately $13.23 million invested during 2012–2015 across 10 research areas (Figure 4-7). The response from USDA reflects investments through its Biotechnology Risk Assessment Grants (BRAG) program, which is jointly administered by USDA's Agricultural Research Service (ARS) and National Institute of Food and Agriculture. The BRAG program receives input regarding its program priorities through multiple regulatory agencies that have an interest in the environmental risk related to the introduction of GE organisms, including USDA–APHIS, EPA, and FDA. About 75 percent of BRAG award recipients for the years included were scientists with land-grant universities or USDA–ARS.

The committee also contacted and received feedback in response to the RFI from several federal agencies that are not primary agencies within the Coordinated Framework. Of the seven

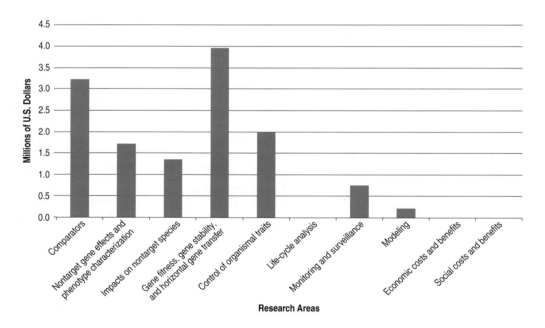

FIGURE 4-7 Spending by the U.S. Department of Agriculture on programmatic work related to fundamental and applied research efforts that can inform human, animal, and ecological risk assessments and socioeconomic costs and benefits, 2012–2015.
NOTES: Life-cycle analysis, economic costs and benefits, and social costs and benefits are outside the authorization of the BRAG program. USDA provided information regarding the awards and which of the research areas an award belongs to. The value entered for projects that fit into more than one research area was obtained by dividing the total value of the award by the number of research areas it fit into.
SOURCE: Information provided by the U.S. Department of Agriculture. Available upon request from the National Academies' Public Access Records Office at PARO@nas.edu.

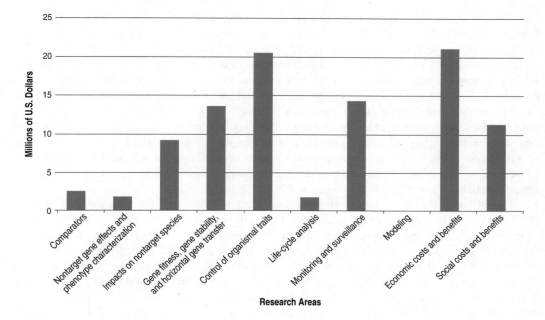

FIGURE 4-8 Spending by the National Science Foundation on programmatic work related to fundamental and applied research efforts that can inform human, animal, and ecological risk assessments and socioeconomic costs and benefits, 2012–2016.
NOTES: SES and CBET submitted information for fiscal years 2012–2015. IIP submitted information for fiscal years 2012–2016. This figure has been updated since the prepublication release. The amount of money directed toward control of organismal traits has been corrected.
SOURCES: Information provided by the National Science Foundation's Division of Industrial Innovation and Partnerships (IIP), Division of Social and Economic Sciences (SES), and Division of Chemical, Bioengineering, Environmental and Transport Systems (CBET). Available upon request from the National Academies' Public Access Records Office at PARO@nas.edu.

offices at the National Science Foundation (NSF) that were sent the RFI, the committee received responses from three of them: the Division of Industrial Innovation and Partnerships, the Division of Social and Economic Sciences, and the Division of Chemical, Bioengineering, Environmental and Transport Systems. These divisions reported total investments of approximately $95.8 million during 2012–2016 (Figure 4-8); additionally, they reported approximately $44.8 million in investments fitting under "other," which primarily involved the product-development research and not research pertaining to the risk assessment of those products. The U.S. Army Corps of Engineers invested $2.08 million during the same time period ($1.24 million in fitness, gene stability, and horizontal gene transfer; and $0.84 million in monitoring and surveillance).

The Office of Naval Research (ONR), the Defense Threat Reduction Agency (DTRA), and the U.S. Army Research Laboratory (ARL) provided the committee with information pertaining to their investments in 2012–2016; most of this research was on biotechnology products (falling under the "other" category) and not directly related to risk-analysis research areas outlined in the RFI. ONR reported approximately $32.9 million, DTRA reported approximately $149.8 million, and ARL reported approximately $88 million in awards pertaining to research and development of biotechnology products.

The Intelligence Advanced Research Projects Agency (IARPA) and the Defense Advanced Research Projects Agency (DARPA) indicated that they were initiating efforts in the areas outline by the RFI as well as research into future products of biotechnology. IARPA indicated to the committee (Julias, 2016) that a 2016 research initiative addressing several of the topics in the RFI was under way and will address comparators; computational modeling; off-target gene effects and phenotypic characterization; fitness, genetic stability, and horizontal gene transfer; control of organismal traits; and impact on nontarget organisms. Although the total level of future investments was not provided, IARPA anticipated the majority of funding would initially address comparators and computational modeling. At the time the committee was writing its report, DARPA was launching a "Safe Genes Program" to support responsible innovation while mitigating the risk of unintended consequences of genome editing and derivative technologies, including gene drives.[17] The level of future research investments for this program was not publicly available.

The Office of Biological and Environmental Research within the U.S. Department of Energy (DOE) indicated investments of approximately $15 million per year between 2012 and 2016 made by the Genomic Science program. The committee was unable to get resolution as to the specific areas (as outlined by the RFI), but it appeared that awards had gone toward the control of organismal traits or future products of biotechnology (the latter falling under the "other" category). Other agencies that responded to the RFI (U.S. Air Force Office of Scientific Research and FDA) indicated that their intramural and extramural programs do not currently address the topics identified by the committee. The committee also heard from the National Invasive Species Council (NISC) Secretariat, which provides support to the NISC to coordinate control of invasive species across the federal government. The council has interest in and has supported work in this area in the past but was not directly funding the areas outlined in the RFI at the time the committee was writing its report.

In summary, for those federal agencies that responded to the RFI, risk-analysis research resources cumulatively totaled to $111.08 million over the 2012–2016 period (Table 4-1). Of note is the lack of any research funding by FDA to address risk analyses for future biotechnology products. Although EPA did not have direct investments for risk analyses of future biotechnology products, its efforts in risk-analysis research in other areas can be applied toward biotechnology products.

A 2015 Woodrow Wilson Center report estimated that in 2008–2014 the U.S. government invested $820 million in synthetic-biology research (with a significant increase in 2010–2014) with DARPA, the U.S. Department of Defense (excluding DARPA), DOE, NSF, and the National Institutes of Health investing the majority resources (Wilson Center, 2015). Of the total investment, the Wilson Center estimated less than 1 percent was invested in risk research and approximately 1 percent was invested in ethical, legal, and social issues. The results of the committee's RFI indicated approximately $6.98 million was invested in social research in 2012–2015, which is fairly consistent with the Wilson Center findings. However, assuming a flat budget for total synthetic-biology research in 2015 as compared to 2014 (that is, approximately $220 million in 2015 or an estimated total of $1.04 billion in 2008–2015), the results of the RFI indicate approximately 9 percent of the total ($89.92 million when excluding social research) was invested in risk–benefit research. Excluding research on economic costs and benefits during the period 2012–2015 ($20.93 million), risk research during that period (approximately $68.98 million) would represent approximately 7 percent of the total synthetic-biology research portfolio.

The results of the Wilson Center study and the committee's RFI indicate that from the "outside looking in" it is difficult to ascertain the level of research funding to support risk analyses, and it appears it may also be a challenge for the U.S. government to aggregate investment totals across agencies. Consequently, the committee acknowledges uncertainty in its estimates of risk-analysis

---

[17]DARPA Safe Genes Proposers Day. Available at https://www.eiseverywhere.com/ehome/196223/443234. Accessed October 11, 2016.

TABLE 4-1 Federal Investment in Risk-Analysis Research, 2012–2015

Type of Risk Research	Amount (in millions)
Comparators	$5.70
Off-target gene effects and phenotypic characterization	$3.02
Impacts on nontarget species	$8.86
Fitness, gene stability, and horizontal gene transfer	$15.44
Control of organismal traits	$19.36
Life-cycle analyses	$1.72
Monitoring and surveillance	$14.67
Modeling	$0.22
Economic costs and benefits	$20.93
Social costs and benefits	$6.98
TOTAL	$96.90

NOTES: This table does not reflect the investments made by IARPA, DARPA, or DOE as there was not sufficient resolution or figures given to be categorized under a specific type of research. A total of $251.27 million was reported to be invested in research that was considered to fall under the "other" category by the committee or the agencies themselves. For consistency, this table represents the amounts reported through 2015, even though some agencies reported figures for 2016. The U.S. Army Corps of Engineers only reported numbers for 2012 and 2015, which are also included here.
SOURCE: Responses to the committee's request for information, available upon request from the National Academies' Public Access Records Office at PARO@nas.edu.

research investments, due the level of resolution different agencies provide for their yearly budgets. The committee also realizes there may be related research efforts outside of synthetic biology that can support future risk-analysis methods (for example, research undertaken by USDA's Agricultural Research Service and the Agricultural Experiment Stations at public universities). However, the estimates suggest that research in specific areas such as modeling and life-cycle analyses, which are critical for supporting premarket risk assessments and socioeconomic cost–benefit analyses, represent 0.02 percent and 0.17 percent, respectively, of the total synthetic-biology research investment. Monitoring and surveillance research, which is critical to supporting post-market assessments and implementing risk-mitigation measures, represents approximately 1.4 percent of the total research portfolio. Research concerning comparators; off-target gene effects and phenotypic characterization; impacts on nontarget species; gene fitness, stability, and horizontal gene transfer; and control of organisms (containment and confinement) combined to represent approximately 5 percent of the total synthetic-biology research investment. These research areas can support both premarket and post-market risk analyses. It is difficult to determine the appropriate level of investment for risk research to support oversight of future biotechnology products; however, the sense of the committee is that the current level is insufficient.

The committee is also concerned that the current U.S. government risk-analysis research portfolio may not be planned in a manner that can maximize its return on investment. It is encouraging that USDA's BRAG program includes other federal agencies, including EPA and FDA, in identifying research priorities; however, there is no indication that these interactions include regulatory risk assessors, risk managers, and researchers working together to vet and adapt research products for use in risk assessments and socioeconomic cost–benefit analyses. Based on the responses to the RFI, there does not appear to be significant interaction between DARPA and other agencies in the U.S. Department of Defense, DOE, and NSF with USDA, EPA, and FDA regulators in research planning or in envisioning a new paradigm for advancing risk-analysis approaches for future bio-technology products.

While the reported research portfolio may be relevant to the risk-analysis needs for future biotechnology products, it is not clear that it is responsive to the nature and extent of future chal-lenges facing public- and private-sector risk assessors, risk managers, and other interested and affected parties. Finally, the committee notes that the financial resources needed to establish an adequate research portfolio for the United States need not fall solely on the U.S. government and the nation's tax payers. The U.S. government may want to explore establishing open and transparent approaches to integrate and optimize public investments, private investments, and public–private partnerships to realize the needed resources to support development of a responsive, nimble, and robust risk-analysis paradigm.

The consequences of current levels of intramural and extramural funding in research and to support future risk-assessment needs suggest the number of products poised to enter the market-place in the coming 5–10 years may outpace the means and capacity for voluntary- or regulatory-based risk–benefit assessments to inform premarket decision making or post-market oversight, including environmental monitoring.

SUMMARY AND CONCLUSIONS

In this chapter, the committee examined whether product risks associated with future biotech-nology will be similar to or different from those of existing biotechnology products and how well understood those risks are. It also reviewed the extent to which the current capabilities of the regu-latory system are appropriately aligned with the likely needs in oversight of those future products. The committee reached the following conclusions.

Conclusion 4-1: The risk-assessment endpoints for future biotechnology products are not new compared with those that have been identified for existing biotechnology products, but the pathways to those endpoints have the potential to be very different in terms of complexity.

The biotechnology products emerging in the next 5–10 years pose a diverse array of envi-ronmental, health, and safety risks that vary widely in terms of their potential impacts, likelihood of occurrence, spatial and temporal dimensions, and the appropriate regulatory policies for their assessment. Although the nature of human health and ecological risk-assessment endpoints that will need consideration are similar to those identified with existing products, the pathways to these end-points will differ in complexity. To the extent future indoor manufacturing will occur in domestic settings, the types of risk-assessment endpoints of future biotechnology products that need to be considered will likely be similar to those used with existing indoor manufacturing; however, the dimensionality of these risk assessments will be more complex and the risk estimates will be more uncertain. Open-release microbial consortia may have more complexity in the dimensionality of their associated risk assessments, but the degree to which that complexity differs from nonengi-neered microorganisms is dependent upon on a number of variables, such as use pattern and envi-ronmental conditions. Such products as well as those designed to suppress, eradicate, or enhance a target species will have new risk pathways and increased dimensionality in the risk assessments.

Transparently elaborated regulatory decisions that provide precedents or comparative examples may shape development strategies for future products. The degree to which risks associated with a product are "well understood" is a subjective determination and formal uncertainty analyses will be needed to determine where a reduced level of regulatory scrutiny may be warranted.

The committee concludes that it is reasonable to assume that existing and future biotechnology products that are similar in terms of their properties, mechanisms, and use patterns may have similar risk profiles, but methods for quantifying similarity between products need to be developed.

Conclusion 4-2: Gaps in the risk-analysis capability of the regulatory system can create a real or perceived impression that some products are entering the marketplace without any government oversight, which can undercut public confidence.

Regulators have the two-fold concern of maintaining and building public confidence in the regulatory process and of engaging in continual improvement of risk analyses to ensure human health and environmental safety. Public confidence in government oversight of emerging technologies may be eroded to the extent there is a lack of transparency and clarity as to how regulatory authorities are undertaking risk assessments, including identifying societal values in addition to taking input from biotechnology developers in formulating regulatory decisions. Given the nature and diversity of future biotechnology products, increased public and developer participation may improve both the understanding and quality of risk-analysis approaches. With better understanding of real or perceived gaps in the risk-analysis process, regulators would be better equipped with the capabilities needed to strengthen risk analysis.

Conclusion 4-3: The expertise and capacity of EPA, FDA, USDA, and other agencies that have interests related to future biotechnology products may not be sufficient to handle the expected scope and scale of future biotechnology products.

Although the regulatory agencies have access to a number of external advisory committees, the number of in-house experts and the responses to the RFI indicate that there may not be sufficient scientific capability, capacity, and tools within and across the agencies to address the risk-assessment challenges for future biotechnology products. The number of products poised to enter the marketplace in the coming years may outpace the means and capacity for voluntary- or regulatory-based assessment processes to inform decision making. This imbalance, if not addressed in the near term, could impede the development of new biotechnology products in the long term. In addition, based on the expected scope and complexity of products of biotechnology that are likely in the next 5–10 years, advances in regulatory science will be needed for effective and appropriate evaluation. Clearly, the profusion of future biotechnology products poses a significant potential stress to the existing regulatory system. Regulatory agencies are likely not prepared with sufficient staff, appropriate risk-analysis approaches, and corresponding guidance for development and evaluation of data packages submitted by product developers.

REFERENCES

Ankley, G.T., R.S. Bennett, R.J. Erickson, D.J. Hoff, M.W. Hornung, R.D. Johnson, D.R. Mount, J.W. Nichols, C.L. Russom, P.K. Schmieder, J.A. Serrano, J.E. Tietge, and D.L. Villeneuve. 2010. Adverse outcome pathways: A conceptual framework to support ecotoxicology research and risk assessment. Environmental Toxicology and Chemistry 29(3):730–741.

DaCunha, C. 2016. Universal Bio Mining: Transforming Mining with Synthetic Biology. Presentation to the National Academies of Sciences, Engineering, and Medicine Committee on Future Biotechnology Products and Opportunities to Enhance Capabilities of the Biotechnology Regulatory System, June 1, Washington, DC.

Drinkwater, K., T. Kuiken, S. Lightfoot, J. McNamara, and K. Oye. 2014. Creating a Research Agenda for the Ecological Implications of Synthetic Biology. Joint Workshops by the MIT Program on Emerging Technologies and the Wilson Center's Synthetic Biology Project. Available at https://www.wilsoncenter.org/sites/default/files/SYNBIO_create%20an%20agenda_v4.pdf. Accessed August 10, 2016.

EFSA (European Food Safety Authority). 2011. Guidance on selection of comparators for the risk assessment of genetically modified plants and derived food and feed. EFSA Journal 9:2149.

EOP (Executive Office of the President). 2017. Modernizing the Regulatory System for Biotechnology Products: An Update to the Coordinated Framework for the Regulation of Biotechnology. Available at https://obamawhitehouse.archives.gov/sites/default/files/microsites/ostp/2017_coordinated_framework_update.pdf. Accessed January 30, 2017.

EPA (U.S. Environmental Protection Agency). 1998. Guidelines for Ecological Risk Assessment. Federal Register 63:26846–26924.

EPA. 2003. Generic Ecological Assessment Endpoints (GEAEs) for Ecological Risk Assessment. EPA/630/P-002/004F. Risk Assessment Forum, Washington, DC. Available at https://www.epa.gov/sites/production/files/2014-11/documents/generic_endpoinsts_2004.pdf. Accessed January 30, 2017.

EPA. 2009. Assessment of Hazard and Exposure Associated with Nanosilver and Other Nanometal Pesticides. FIFRA Scientific Advisory Panel Meeting, November 3–6. Available at https://www.regulations.gov/docket?D=EPA-HQ-OPP-2009-0683. Accessed January 14, 2017.

EPA. 2011. Integrated Approaches to Testing and Assessment Strategies (IATA): Use of New Computational and Molecular Tools. FIFRA SAP Meeting May 24-26, 2011. Available at https://www.regulations.gov/docket?D=EPA-HQ-OPP-2011-0284. Accessed October 25, 2016.

EPA. 2012. FIFRA SAP Meeting on Pollinator Risk Assessment Framework, September 11–16. Available at https://www.regulations.gov/docket?D=EPA-HQ-OPP-2012-0543. Accessed January 14, 2017.

EPA. 2014a. Notice of FIFRA SAP Meeting; RNAi Technology as a Pesticide: Problem Formulation for Human Health and Ecological Risk Assessment, January 28. Available at https://www.regulations.gov/docket?D=EPA-HQ-OPP-2013-0485. Accessed on August 10, 2016.

EPA. 2014b. Risk Assessment Forum White Paper: Probabilistic Risk Assessment Methods and Case Studies. EPA/100/R-14/004. Risk Assessment Forum, Washington, DC. Available at https://www.epa.gov/sites/production/files/2014-12/documents/raf-pra-white-paper-final.pdf. Accessed on November 20, 2016.

EPA. 2015. Development of a Spatial Aquatic Model (SAM) for Pesticide Risk Assessment. FIFRA Scientific Advisory Panel Meeting, September 15–18. Available at https://www.regulations.gov/docket?D=EPA-HQ-OPP-2015-0424. Accessed January 14, 2017.

EPA. 2016a. DEEM-FCID/Calendex Software Installer. Available at https://www.epa.gov/pesticide-science-and-assessing-pesticide-risks/deem-fcidcalendex-software-installer. Accessed on August 10, 2016.

EPA. 2016b. Public Participation Process for Registration Actions. Available at https://www.epa.gov/pesticide-registration/public-participation-process-registration-actions. Accessed on August 10, 2016.

EPA. 2016c. FIFRA Scientific Advisory Panel; Notice of Public Meeting. Available at https://www.regulations.gov/docket?D=EPA-HQ-OPP-2013-0485. Accessed on August 10, 2016.

EPA. 2016d. EPA Sustainable and Healthy Communities National Research Program: 2015 Accomplishments. EPA 601/R-16/004. Office of Research and Development, EPA. Available at https://www.epa.gov/sites/production/files/2016-08/documents/shc2015finalepa601-r16-004web-508_metadata.pdf. Accessed September 13, 2016.

EPA. 2016e. TIM Version 3.0 Beta–Technical Description and User's Guidance. Available at https://www.epa.gov/pesticide-science-and-assessing-pesticide-risks/tim-version-30-beta-technical-description-and-users. Accessed August 10, 2016.

EPA. 2016f. Research on Analyzing the Life Cycle of Chemicals. Available at https://www.epa.gov/chemical-research/research-analyzing-life-cycle-chemicals. Accessed August 10, 2016.

EPA. 2016g. Generic Ecological Assessment Endpoints (GEAEs) for Ecological Risk Assessment: Second Edition with Generic Ecosystem Services Endpoints Added. EPA/100/F15/005. Risk Assessment Forum, Washington, DC. Available at https://www.epa.gov/sites/production/files/2016-08/documents/geae_2nd_edition.pdf. Accessed November 20, 2016.

Fairbrother, A., J. Gentile, C. Menzie, and W. Munns. 1999. Report on the Shrimp Virus Peer Review and Risk Assessment Workshop: Developing a Qualitative Risk Assessment. Washington, DC: EPA.

FDA (U.S. Food and Drug Administration). 2016. Preliminary Finding of No Significant Impact in Support of an Investigational Field Trial of OX513A *Aedes aegypti* Mosquitoes. March. Available at http://www.fda.gov/downloads/AnimalVeterinary/DevelopmentApprovalProcess/GeneticEngineering/GeneticallyEngineeredAnimals/UCM487379.pdf. Accessed September 30, 2016.

Focazio, M.J., D.W. Kolpin, K.K. Barnes, E.T. Furlong, M.T. Meyer, S.D. Zaugg, L.B. Barber, and M.E. Thurman. 2008. A national reconnaissance for pharmaceuticals and other organic wastewater contaminants in the United States — II. Untreated drinking water sources. Science of the Total Environment 402(2-3):201–216.

Funtowicz, S.O., and J.R. Ravetz. 1995. Science for the post normal age. Pp. 146–161 in Perspectives on Ecological Integrity, L. Westra and J. Lemons, eds. Dordrecht, Netherlands: Springer.

Gallagher, S.G., G.E. Rice, L.J. Scarano, L.K. Teuschler, G. Bollweg, and L. Martin. 2015. Cumulative risk assessment lessons learned: A review of case studies and issue papers. Chemosphere 120:697–705.

Greenblott, J.M., T. O'Farrell, and B. Burchard. 2016. EPA's strategic foresight pilot project. Presentation at the Institute for Alternative Futures' Public Sector Foresight Network and Federal Foresight Community of Interest Foresight Day, July 22, Washington, DC.

IOM (Institute of Medicine). 2013. Environmental Decisions in the Face of Uncertainty. Washington, DC: The National Academies Press.

Julias, J. 2016. IARPA Overview of NAS RFI. Webinar presentation to the National Academies of Sciences, Engineering, and Medicine Committee on Future Biotechnology Products and Opportunities to Enhance Capabilities of the Biotechnology Regulatory System, July 25.

Kuzma, J., J. Larson, and P. Najmaie. 2009a. Evaluating oversight systems for emerging technologies: A case study of genetically engineered organisms. Journal of Law, Medicine, and Ethics 37(4):546–586.

Kuzma, J., A. Kuzhabekova, and K. Wilder. 2009b. Improving oversight of genetically engineered organisms. Policy and Society 28(4):279–299.

Kuzma, J., A. Kuzhabekova, S. Priest, and L. Yerhot. 2010. Expert opinion of emerging technologies oversight: Lessons for nanotechnology from biotechnology. Pp. 133–156 in Understanding Nanotechnology: Philosophy, Policy, and Publics, U. Fiedeler, C. Coenen, S.R. Davies, and A. Ferrari, eds. Amsterdam: IOS Press.

Landis, W. 2004. Ecological risk assessment conceptual model formulation for nonindigenous species. Risk Analysis 24(4):847–858.

Murphy, B., C. Jansen, J. Murray, and P. De Barro. 2010. Risk Analysis on the Australian Release of *Aedes aegypti* (L.) (Diptera: Culicidae) Containing *Wolbachia*. Indooroopilly, Australia: CSIRO.

Murray, J.V., C.C. Jansen, and P. De Barro. 2016. Risk associated with the release of *Wolbachia*-infected *Aedes aegypti* mosquitoes into the environment in an effort to control dengue. Frontiers in Public Health 4:43.

NAPA (National Academy of Public Administration). 2016. Strategic foresight. Pp. 7–11 in Transition 2016: Equipping the Government for Success in 2016 and Beyond. Available at http://www.napat16.org/images/7.8.2016_T16_Final_Report.pdf. Accessed February 12, 2017.

NAS (National Academy of Sciences). 1987. Introduction of Recombinant DNA-Engineered Organisms into the Environment: Key Issues. Washington, DC: National Academy Press.

NASEM (National Academies of Sciences, Engineering, and Medicine). 2016a. Gene Drives on the Horizon: Advancing Science, Navigating Uncertainty, and Aligning Research with Public Values. Washington, DC: The National Academies Press.

NASEM. 2016b. Genetically Engineered Crops: Experiences and Prospects. Washington, DC: The National Academies Press.

Nelson, K.C., G. Kibata, L. Muhammad, J.O. Okuro, F. Muyekho, M. Odindo, A. Ely, and J.M. Waquil. 2004. Problem formulation and options assessment (PFOA) for genetically modified organisms: The Kenya case study. Pp. 57–82 in Environmental Risk Assessment of Genetically Modified Organisms, Vol. 1. A Case Study of *Bt* Maize in Kenya, A. Hilbeck and D. Andow, eds. Cambridge, MA: CABI Publishing.

NRC (National Research Council). 1996. Understanding Risk: Informing Decisions in a Democratic Society. Washington, DC: National Academy Press.

NRC. 2000. Genetically Modified Pest-Protected Plants: Science and Regulation. Washington, DC: National Academy Press.

NRC. 2007. Toxicity Testing in the 21st Century: A Vision and a Strategy. Washington, DC: The National Academies Press.

NRC. 2008. Public Participation in Environmental Assessment and Decision Making. Washington, DC: The National Academies Press.

NRC. 2009. Science and Decisions: Advancing Risk Assessment. Washington, DC: The National Academies Press.

NRC. 2012. Science for Environmental Protection: The Road Ahead. Washington, DC: The National Academies Press.

NRC. 2013. Assessing Risks to Endangered and Threatened Species from Pesticides. Washington, DC: The National Academies Press.

NSTC (National Science and Technology Council). 2007. Agricultural Biotechnology Risk Analysis Research in the Federal Government: Cross Agency Cooperation. Available at https://www.nsf.gov/pubs/2007/nsf07208/nsf07208.pdf. Accessed October 26, 2016.

Orr, R. 2003. Generic nonindigenous aquatic organism risk analysis. Pp. 415–438 in Invasive Species: Vectors and Management Practices, G.M. Ruiz and J.T. Carlton, eds. Washington, DC: Island Press.

Phillips, P.J., C. Schubert, D. Argue, I. Fisher, E.T. Furlong, W. Foreman, J. Gray, and A. Chalmers. 2015. Concentrations of hormones, pharmaceuticals and other micropollutants in groundwater affected by septic systems in New England and New York. Science of the Total Environment 512–513:43–54.

Reardon, S. 2016. Welcome to the CRISPR zoo. Nature 431(7593):160–163.

Renn, O. 1992. Concepts of risk: A classification. Pp. 53–79 in Social Theories of Risk, S. Krimsky and D. Golding, eds. Westport, CT: Praeger.

Renn, O. 2005. Risk Governance: Towards an Integrative Approach. Geneva: International Risk Governance Council.

Ricroch, A.E., and M.-C. Hénard-Damave. 2016. Next biotech plants: New traits, crops, developers and technologies for addressing global challenges. Critical Reviews in Biotechnology 36(4):675–690.

Roberts, J.P., S. Stauffer, C. Cummings, and J. Kuzma. 2015. Synthetic Biology Governance: Delphi Study Workshop Report. GES Center Report No. 2015.2. Available at https://research.ncsu.edu/ges/files/2014/04/Sloan-Workshop-Report-final-ss-081315-1.pdf. Accessed October 10, 2016.

Rose, R., S. McCammon, and S. Lively. 2006. Proceedings: Workshop on Confinement of Genetically Engineered Crops During Field Testing. U.S. Department of Agriculture. Available at https://www.aphis.usda.gov/brs/pdf/conf_ws_proc2.pdf. Accessed October 4, 2016.

Schaider, L.A., R.A. Rudel, J.M. Ackerman, S.C. Dunagan, and J.G. Brody. 2014. Pharmaceuticals, perfluorosurfactants, and other organic wasterwater compounds in public drinking water wells in a shallow sand and gravel aquifer. Science of the Total Environment 468–469:384–393.

Selin, C., K.C. Rawlings, K. de Ridder-Vignone, J. Sadowski, C.A. Allende, G. Gano, S.R. Davies, and D.H. Guston. 2016. Experiments in engagement: Designing public engagement with science and technology for capacity building. Public Understanding of Science [online]. Available at http://pus.sagepub.com/content/early/2016/01/13/0963662515620970.full.pdf+html. Accessed October 25, 2016.

SEP (Science for Environment Policy). 2016. Future Brief: Identifying Emerging Risks for Environmental Policies. Issue 13. Bristol, UK: European Commission.

SRA (Society for Risk Analysis). 2015. SRA glossary. June 22. Available at http://sra.org/sites/default/files/pdf/SRA-glossary-approved22june2015-x.pdf. Accessed January 14, 2017.

Sunstein, C.R. 2005. Laws of Fear: Beyond the Precautionary Principle. Cambridge, UK: Cambridge University Press.

Suter, G.W. 2007. Ecological Risk Assessment. Boca Raton, FL: CRC Press.

Traynor, P.L., and J.H. Westwood, eds. 1999. Proceedings of a Workshop on Ecological Effects of Pest Resistance Genes in Managed Ecosystems, January 31–February 3, Bethesda, MD. Available at http://www.isb.vt.edu/documents/proceedings.pdf. Accessed October 4, 2016.

USDA–APHIS (U.S. Department of Agriculture–Animal and Plant Health Inspection Service). 2004. Environmental Impact Statement; Introduction of Genetically Engineered Organisms. Federal Register 69:3271–3271.

USDA–APHIS. 2016. Environmental Impact Statement; Introduction of the Products of Biotechnology. Federal Register 81:6225–6229.

Warren-Hicks, W.J., and A. Hart, eds. 2010. Application of Uncertainty Analysis to Ecological Risks of Pesticides. Boca Raton, FL: CRC Press.

Wilson Center. 2015. U.S. Trends in Synthetic Biology Research Funding. Synthetic Biology Project. Washington, DC: Woodrow Wilson Center for International Scholars.

Wolt, J.D., and R.K.D. Peterson. 2010. Prospective formulation of environmental risk assessments: Probabilistic screening for Cry1A(b) maize risk to aquatic insects. Ecotoxicology and Environmental Safety 73:1182–1188.

Writer, J.H., I. Ferrer, L.B. Barber, and E.M. Thurman. 2013. Widespread occurrence of neuro-active pharmaceuticals and metabolites in 24 Minnesota rivers and wastewaters. Science of the Total Environment 461–462:519–527.

Yeardley, R.B., Jr., B. Dyson, and M. Tenbrink. 2011. EPA Growing DASEES (Decision Analysis System for a Sustainable Environment, Economy and Society) To Aid in Making Decisions on Complex Environmental Issues. EPA/600/F-11/023. Washington, DC: EPA. Available at https://cfpub.epa.gov/si/si_public_record_report.cfm?dirEntryId=238232. Accessed September 24, 2016.

5

Opportunities to Enhance the Capabilities of the Biotechnology Regulatory System

The profusion of products and the growing number of actors in the biotechnology space described in Chapter 2 present many challenges to the U.S. biotechnology regulatory system. The present chapter outlines a framework for risk analysis targeted at the types and scale of products anticipated and describes what tools, expertise, and scientific capabilities are required within and beyond the regulatory agencies in order to support oversight of future biotechnology products. The focus is not just on the regulatory process, but the broader context of presubmission and post-market activities that are an important part of the overall regulatory framework and that can provide a balanced approach to capabilities required for regulation of future biotechnology products.

As technologies and basic knowledge advance, a regulatory system should be able to adapt to new risks of future biotechnology products and also to adjust to well-established categories of products as their level and types of risk become better understood. As discussed in Chapters 2, 3, and 4, the scope, scale, complexity, and tempo of future products is expected to increase rapidly, and this increase has the potential to overwhelm the existing regulatory system. In addition, the new types of actors and new types of business models that will be involved in the development of technology and products means that the regulatory system will likely need to provide information to a broader group of stakeholders with diverse backgrounds and expertise. Finally, the possibility that some future products of biotechnology will be controversial may require substantial conversation and public debate throughout the phases of the regulatory process. A regulatory system with a greater emphasis on stratified approaches that prioritize the regulatory agencies' familiarity with a product, the complexity of the risk assessment for the product, and the anticipated risk associated with the product (that is, proportionate oversight) could contribute to meeting the increased demands on the system.

The 2016 *National Strategy for Modernizing the Regulatory System for Biotechnology Products* issued by the Executive Office of the President recognizes the increased complexity of future biotechnology products and provides a strategic plan for ensuring that federal agencies can efficiently assess any risks associated with such products (EOP, 2016). It also describes several approaches to increasing public participation in the process and incorporating science-driven decision making

BOX 5-1
Governance, Oversight, and Regulation

The terms governance, oversight, and regulation are used in regulatory science to capture different aspects of risk management (Figure 5-1). Following Kuzma (2006), *governance* can be broadly defined as a complex set of values, norms, processes, and institutions through which society manages technology development and deployment and resolves conflict formally or informally. Governance includes *oversight*, which is defined more narrowly as watchful and responsible care or regulatory supervision. *Regulation* is a subcategory of oversight and governance and represents an authoritative rule dealing with details or procedure or a rule or order issued by an executive authority or regulatory agency of a government and having the force of law. Therefore, regulation can be an important element of governance but can also be excluded from a governance system. Oversight can include codes of conduct, voluntary data-sharing programs, and public–private partnerships for certification standards as well as regulations. Risk analysis for future products of biotechnology can occur within a formal statute-based regulatory system or outside of one. Governance can include risk analyses from standard-setting international bodies, academics, nongovernmental organizations, think-tanks, and companies, whether or not those products are submitted for formal regulatory oversight. Oversight programs like voluntary standard setting, sharing of data, and risk-mitigation activities can occur outside of legal authorities.

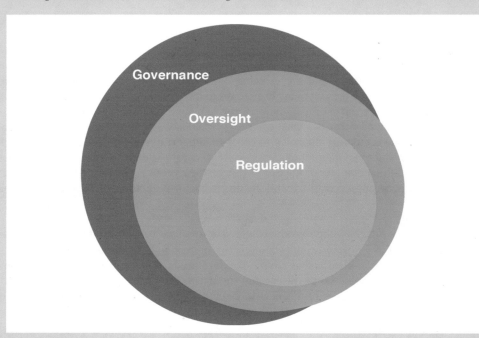

FIGURE 5-1 Relationships of governance, oversight, and regulation.
SOURCE: Illustration by J. Kuzma.

(EOP, 2016). This chapter describes some of the properties that will be important for risk analysis of the next generation of products, with the intent of providing insight that can be used by the agencies in evaluating the capabilities required to perform appropriate oversight (Box 5-1).

CONSISTENT, EFFICIENT, AND EFFECTIVE DECISION MAKING FOR FUTURE PRODUCTS OF BIOTECHNOLOGY

A key property of the U.S. biotechnology regulatory system, well articulated in the update to the Coordinated Framework, is that it effectively protects human health and the environment through a safety-evaluation process that can be understood by members of the public (EOP, 2017). As described in Chapter 3, the structure of the Coordinated Framework presents considerable flexibility for regulating future products of biotechnology but requires the agencies to appropriately apply their statutory authority. Multiple agencies may have jurisdiction over a given product, while other products may not be explicitly covered by any statute or federal agency. Either situation could lead to uncertainty in regulatory jurisdiction. The regulatory route may also be unclear at the time future products are developed. This uncertainty in regulatory jurisdiction "can make it difficult for the public to understand how the safety of biotechnology products is evaluated and create challenges for small and mid-sized businesses navigating the regulatory process" (EOP, 2017:1). As articulated in both the update to the Coordinated Framework and the *National Strategy*, the desired state is one in which there is a consistent, efficient, and effective decision-making framework that continues to protect human health and the environment. This section provides some properties of the risk-analysis system that will be important for meeting these goals for products anticipated in the next 5–10 years.

Preparing for Increased Scope, Scale, and Complexity of Biotechnology Products

A key theme throughout this report is the increase in scope, scale, and complexity that will accompany future biotechnology products. Scope is the new types of biotechnology products that have not yet been seen by regulators. Scale refers to the number of products as well as the number of variants of products that may interact with the regulatory system. Complexity refers to the number of traits that may be involved in a single product and the interactions between the various elements in a product. Increased scope and complexity are key components of future products that may have fewer or no comparators to nonbiotechnology products or no similar existing biotechnology products and thus little or no familiarity within the regulatory system.

Though the scale of products is likely to increase, some of this volume will be comprised of new products with a composition similar to existing biotechnology products with a history of characterization and safe use. Such products should be familiar to regulatory agencies and should have low complexity because the risk analyses for such products are well understood. The introduction of an already approved *Bt* protein into a new crop variety is an example. Another example of a product for which an *a priori* argument for familiarity and low complexity might be made is an organism that contains only a loss-of-function mutation in a gene or genes because such mutations arise spontaneously in nature. Provided the loss-of-function mutation does not create a new reading frame that encodes a novel protein, an organism with such a mutation is likely to be not complex in terms of risk analysis. A benefit of products that are *familiar and not complex* is the savings to regulators in terms of time and effort spent on designing and implementing risk analyses. These savings in time and resources can then be applied to devising and implementing risk analyses for products that are less familiar, more complex, or less familiar and more complex. It will be important in implementing the update to the Coordinated Framework (EOP, 2017) to make use of scientific tools to evaluate when new products can be categorized as *familiar and not complex* by comparison with the existing base of scientific knowledge and to apply appropriate oversight to those products (including no regulatory oversight, if appropriate) based on scientifically sound risk analyses.

Other new products—such as organisms with entirely new pathways assembled from many genes derived from multiple unrelated sources, perhaps including synthetic genes, and engineered microbial communities planned for open-environmental release in which some community members contain engineered pathways—will pose challenges for the regulatory system because the regulatory agencies have not seen these types of products before and because the products do not have nonbiotechnology products to which they can be easily compared. Such products would be *unfamiliar* and have *high complexity* for the regulatory agencies.

Examples of products that pose new regulatory challenges are organisms engineered to contain gene drives, which are designed to introduce a trait that spreads throughout the species population. A trait could be designed to modify a species, such as one that reduces the species' ability to transmit a disease, or to eliminate a species, which may be the case when trying to exterminate a particular disease vector from a geographical region. In the case of gene drives in insects, the same public health benefit of disease elimination could be attempted by releasing sterile males of the species (Krafsur, 1998; Benedict and Robinson, 2003), but the use of a gene drive may be more effective in reducing the population size of the target species. However, gene drives may pose new complexity for risk assessments if the speed with which the target-species population is depressed exceeds current ecological and evolutionary rates. Additional risk-assessment endpoints and pathways to those endpoints may also need to be addressed. Examples of pathways to risk-assessment endpoints could include the probability that off-target gene effects could result in an unanticipated phenotype, the probability that the gene drive could mutate and result in an unanticipated phenotype, or the changes that the system (or its mutations) could cause in a community food web. Although these examples do not represent new risk-assessment endpoints, they may require more sophisticated risk analyses, with consideration of increasingly complex interactions. As noted in Recommendation 6-3 of the National Academies report on gene drives (NASEM, 2016a:128): "To facilitate appropriate interpretation of the outcomes of an ecological risk assessment, researchers and risk assessors should collaborate early and often to design studies that will provide the information needed to evaluate risks of gene drives and reduce uncertainty to the extent possible."

Another example of a new type of product is one that would enable the "deextinction" of a species. At the time the committee was writing its report, there were projects under way to "deextinct" the passenger pigeon and the woolly mammoth (or arguably a relative), among other animals (Biello, 2014; Callaway, 2015; Shapiro, 2015). If release to a natural ecosystem is a goal of such a project, a meaningful risk assessment should include wildlife ecologists and local experts from the area of release, including those with knowledge about migratory routes, to assist in assessing effects on the existing function and structure in the community.

An important approach for dealing with an increase in the products of biotechnology will be the increased use of stratified approaches to regulation, where new and potentially more complex risk-analysis methods will need to be developed for some products, while established risk-analysis methods can be applied or modified to address products that are familiar or that require less complex risk analysis. With this approach, new risk-analysis methods are focused on products with less familiar characteristics, more complex risk pathways, or both. Multiple criteria are usually embedded within risk analyses to ascertain if an estimated level of risk is consistent with the risk-management goals established during the problem-formulation phase of a risk assessment. In some cases, additional risk analyses may be needed to refine risk estimates, to evaluate risk-mitigation measures, or both. Criteria that could be applied to biotechnology products have also been used for risk analysis of other emerging technologies that integrate health, environmental, and life-cycle effects and occupational and socioeconomic risks, and these criteria can be weighted and rated by experts or stakeholders (Linkov et al., 2007; Tsang et al., 2014). In order to implement the appropriate rigor of risk analyses for new biotechnology products, it will be necessary to establish scientifically rigorous criteria based on factors affecting the perception of risk, the degree of uncertainty, and the magnitude of risk and nature of potential risks.

Enhancing the Responsiveness of the Regulatory System

At the time the committee was writing its report, there was no regulation, law, or statute to mandate a central review of biotechnology products or to develop an oversight system that is coordinated among agencies, minimizes gaps and redundancies in product review, provides more certainty for product developers as to the regulatory path, and embraces the principles of anticipation, participation, responsiveness, and transparency. The update to the Coordinated Framework (EOP, 2017) and the *National Strategy* (EOP, 2016) recognize the need for addressing these issues and provide a set of first steps for doing so. In this section, the committee provides some insights on how these topics might be addressed for the types of products that are anticipated in the next 5–10 years.

As described in Chapter 3, the statutory authorities that apply to some of the future products of biotechnology can be confusing and better coordination among the agencies would be beneficial so that risk analyses cover the impacts of biotechnology products more comprehensively in some cases or avoid duplication of data submissions in others. For example, as of 2016, genetically engineered insects were regulated by the U.S. Food and Drug Administration (FDA) under the Federal Food, Drug, and Cosmetic Act (FDCA) and environmental assessments were performed under the National Environmental Protection Act (NEPA).[1] Crops with resistance to targeted insects through the insertion of genetic material from *Bacillus thuringiensis* were reviewed by the U.S. Department of Agriculture (USDA; under the Plant Protection Act to evaluate if the crop could be a pest) and the U.S. Environmental Protection Agency (EPA; under the Federal Insecticide, Fungicide, and Rodenticide Act [FIFRA]) to determine if the *Bt* toxin, the insecticide produced by the plant, would cause unreasonable adverse effects to humans and the environment); product developers of these crops also consulted with FDA (under the FDCA) before commercial release to ensure the food products derived from the engineered plant were substantially equivalent to corn products already in the marketplace. These examples underscore that developers of future products of biotechnology would benefit substantially from access to timely, consistent, and unambiguous feedback from the federal regulatory system as to whether or not a product is regulated and, if so, which agency or agencies would be response for regulatory oversight.

One possible approach would be to consider the development of a single "point of entry" as a mechanism for initiating the cross-agency cooperation that is articulated in the update to the Coordinated Framework and in the *National Strategy*. Box 5-2 provides an example of what such a mechanism could look like that would operate with the agencies' existing statutory authorities. A collection of integrated resources could be maintained that provide a means for developers to initially determine if their product falls under regulation and, if so, an initial "read" on the regulatory pathway likely to be required for a future regulatory decision. A single point of entry could also provide an accessible public face for the regulatory system where interested parties can explore and understand the nature of the regulatory process. In addition, such a point of entry could be used to enable the federal agencies to decide early in the product-development cycle which authorities are relevant in cases where there have not been precedents. Throughout the process, developers would also have access to ombudsmen within each agency for additional assistance and feedback, including an opportunity to meet with the lead agency prior to a decision on a proposed oversight approach.

The concept of a single point of entry is already available for some distinct parts of the regulatory system; for example, crop developers can submit a letter to the "Am I Regulated" site[2] of

---

[1]In January 2017, FDA issued a draft guidance on mosquito-related products to clarify that its definition of nonfood regulated articles no longer included those "intended to function as pesticides by preventing, destroying, repelling, or mitigating mosquitoes for population control purposes. FDA believes that this interpretation is consistent with congressional intent and provides a rational approach for dividing responsibilities between FDA and EPA in regulating mosquito-related products" (FDA, 2017:6575).

[2]Am I Regulated Under 7 CFR part 340? Available at https://www.aphis.usda.gov/aphis/ourfocus/biotechnology/am-i-regulated. Accessed January 15, 2017.

BOX 5-2
Use of a Single Point of Entry for Application of Risk
Analysis to Future Products of Biotechnology

Figure 5-2 describes a possible structure for providing a stratified approach to regulatory assessment of future products of biotechnology and explains how this structure could be used in a larger risk-assessment framework as an illustration of how a variety of science-based mechanisms might be useful in considering future products of biotechnology.

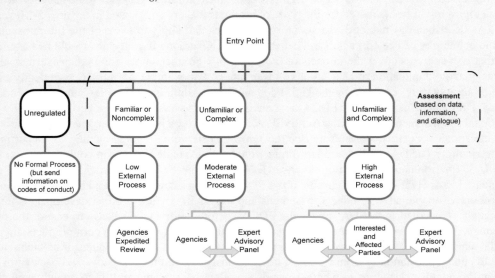

FIGURE 5-2 Providing access to the U.S. regulatory system through a single point of entry.
NOTES: Potential product developers and interested parties would begin by going to an entry point and providing characteristics of the intended product and its use pattern. If the product does not fall under a federal statute, the developer would be notified the product is not federally regulated. If the product is regulated, the appropriate agency or agencies would be identified for the developer. An evaluation of the product's familiarity to regulatory agencies and its complexity in terms of risk analyses as compared to existing biotechnology products would be ascertained (red bins). Depending on the product's familiarity and the complexity of its risk analysis, a different set of risk-analysis processes would be employed (blue boxes). For products that are *familiar* to the regulatory agencies and are *not complex*, a more expedited process could be used, under the assumption that relevant risk-analysis processes are well established. For products that are *less familiar, more complex,* or *less familiar and more complex,* increasingly unique risk-analysis processes (that incorporate additional external input) may need to be established.

Starting at the top of the diagram, a first query from a product developer might be to evaluate whether the intended use of the product is regulated under a given statute. Developers and their advisors (for example, legal counsel and risk-analyses consultants) can independently decide if their products are or are not regulated, but the single point of entry provides a voluntary opportunity to get input from the regulatory agencies. If the product is not regulated (the "unregulated" bin), the resources available through the point of entry could also provide information about voluntary stewardship programs if available.[a] If the intended product is regulated, the developer would be informed as to which regulatory agency or agencies have oversight. A determination would also be made as to whether the product is *familiar and not complex,* is *unfamiliar or complex,* or is *unfamiliar and complex* compared to existing biotechnology products. Products would be assumed to have increasing levels of uncertainty in risk estimates depending on the level of familiarity and the degree of complexity, but the product's actual probability of causing adverse effects, once determined, may or may not be of concern, based on the statutory requirements relevant for the product's use pattern. For products that are *unfamiliar and complex,* this process could involve external input from stakeholders and experts.

Once a determination has been made, the appropriate processes within the relevant agency (or agencies) would be used to provide the necessary risk analysis to support a regulatory decision. Products that fall in the bin of *familiar and noncomplex* could be regulated by notification (for example, some FDA and USDA actions) or a rapid, streamlined approval (for example, EPA's 90-day decision time frame for new biotechnology organisms under the Toxic Substances Control Act and its streamlined risk analyses for substantially similar pesticide products under the Federal Insecticide, Fungicide, and Rodenticide Act). Products for open release into minimally managed or unmanaged environments might be covered under a programmatic NEPA finding of no significant impact or environmental assessment, an Endangered Species Act no-effects determination or consultation, or both, as appropriate. Products that are determined to be *unfamiliar or complex* or *unfamiliar and complex* would likely require modification of or establishment of risk-analysis methods because there would be little or no existing regulatory decisions from which existing risk analyses could be directly applied. The level of effort needed to develop risk analyses would increase moving from the *unfamiliar or complex* bin to the *unfamiliar and complex* bin. There could also be additional risk-assessment tiers for some products within the bins depending on the pattern of the criteria mentioned in the text. The amount and nature of external input (see NRC, 1996, 2008, 2009) would depend on a product's level of familiarity and the degree of complexity. Assessments would also be expected for new use-pattern requests of existing products; however, depending on the similarity to an existing approved use, the level of effort for all parties could be reduced.

Following a decision, intensity of post-market surveillance or monitoring (if on a case-by-case basis it is determined to be necessary to address a risk-assessment uncertainty or assess effectiveness of a risk-mitigation measure) would be scaled with the outcome of the risk-based regulatory decision. Monitoring would likely be more intensive for open-release products associated with the *unfamiliar and complex* bin. Depending on the type of product, some may be required to undergo statutorily mandated reevaluation in specified time frames or as needed based on results of monitoring information.

A desirable feature of an integrated, stratified approach to regulatory oversight is that over time product types originally placed in the *unfamiliar or complex* bin or the *unfamiliar and complex* bin would "move" to a bin of less complexity or more familiarity based on experience gained in evaluating additional products in a category. This paradigm does not imply that products that are *familiar and not complex* necessarily have a low probability of causing adverse effects nor does it imply a product with less familiarity or more complexity necessarily has a high probability of causing adverse effects. Rather, for products that are *familiar and not complex*, the developer's and agency's risk assessors and managers and interested and affected parties can draw upon existing information and risk analyses for similar products, which should facilitate the efficiency of the regulatory decision even if a complex risk analysis is required. For products that are *unfamiliar or complex* or *unfamiliar and complex*, the risk-analysis processes may need to be developed based on limited information and experience and may perhaps require a *de novo* approach. These risk-analysis approaches would likely benefit from external scientific peer review and input from interested and affected parties. Regardless of the initial determination, risk-analysis approaches for products may become more or less complicated over time as new information from monitoring or additional laboratory and field studies becomes available. Proposed decisions to move product types between bins could include public comment and could be informed by external peer review, using best available science, and external party engagement as appropriate.

The outcome of external peer reviews of products evaluated through this process could also help inform the agencies' research agenda to support risk-assessment and risk-management decision making. In this regard the process is envisioned to reflect a design-build-test-learn paradigm in the development and application of risk-based decision making. In addition, developers for products in the *unfamiliar or complex* bin or the *unfamiliar and complex* bin could be encouraged to engage with the appropriate authorities early and while the product is still in the research and development pipeline to help guide dialogue on information needs for the assessment and streamline or target information needs for risk assessments.

While this approach does not eliminate the time and resource investments for a developer pursuing a first-of-a-kind product, data compensation measures in existing statutes, reduced registration fees for small business, and assistance grants from the small business administration, for example, could reduce the financial burden for smaller companies.

---

[a]For products that do not fall under a regulatory authority, industry or nongovernmental organization consortia could develop stewardship programs or third-party certification procedures that, as appropriate, mimic principles in the proposed framework.

USDA's Animal and Plant Health Inspection Service (APHIS) to find out if the agency considers their crop a regulated article. This process lets the crop developers know earlier if their crop is regulated or not, and it lets USDA know earlier what kind of crops are being developed. The concept is also a stated intent of the *National Strategy*. Descriptions were given in the *National Strategy* for multiple online resources maintained by each of EPA, FDA, and USDA, though these were not yet integrated at the time the committee's report was written and hence product developers and other interested parties had to navigate multiple sites that reflect the complexity of the regulatory system and the agencies' jurisdictions. There are examples from the European Union that collect together various product types into a single point of entry and provide a means for public consultation in the context of allergenicity assessment.[3] A similar system for the U.S. regulatory system could provide a more easily navigated system for identifying the regulatory routes for a given product class.

It was not within the committee's statement of task to delineate how a single point of entry could be crafted and implemented. As mentioned, such a mechanism could operate within the agencies' statutory authorities and could range in concept from greater cooperation among the agencies in terms of sharing resources to more consistent and rigorous interagency working group collaboration. Alternatively, it could be operated by an existing coordinating unit within the executive branch or by a new agency created to be the "front door" for all biotechnology products, although the latter option would require new legislation from Congress. However it might be constructed, a key element of an effective single point of entry will be the establishment of criteria that provide guidance on the regulatory route that will be required. This guidance would not necessarily be exclusively consultative or structured through case-by-case deliberations. There are good examples of published guidance used within federal agencies that provide interested parties with relevant information, such as the content of agency website information regarding navigation through the system, and methodological guidance, such as EPA Guidelines for Ecological Risk Assessment (EPA, 1998) and FDA Guidance for Industry.[4] There are clear needs for this information to be improved and continually updated, and this would be an important facet of the point-of-entry implementation. Internationally, regulatory guidance is more commonly available than in the United States; examples are the European Food Safety Authority guidance developed in response to European Union directives (EFSA, 2010, 2011a,b,c). Experience elsewhere with the use of such guidance could be considered when designing a single point of entry to be used within the Coordinated Framework.

As described in Box 5-2, the criteria for which bin a product would fall into would be based on familiarity with existing, regulated products (there should be greater certainty as to how to undertake a risk assessment with a familiar product as compared to an unfamiliar product). Additional product attributes such as the degree of confinement and/or containment (greater confinement/containment should reduce the likelihood of environmental exposure), whether it is living or nonliving (a living product may increase uncertainty and unpredictability of the assessment), and reversible or nonreversible product deployment (a nonreversible deployment may increase the complexity of risk-management measures to mitigate adverse effects) need to be considered in determining the appropriate bin for a new product (see Figure 5-2). The greater the amount and specificity of information a developer can provide for a product (including a proposed risk-analysis approach) through the single entry point, the more efficiently the agencies should be able to determine the product's level of familiarity and the degree of complexity. The development and use of the multidimensional decision criteria for bin placement could be informed by external, independent peer review and input from interested and affected parties. Developers might be able to self-score their product as to the appropriate bin, but the ultimate determination would be an inherently govern-

---

[3]Register of Questions. Available at http://registerofquestions.efsa.europa.eu/roqFrontend. Accessed January 15, 2017.

[4]Guidance for Industry, Biotechnology Guidances. Available at http://www.fda.gov/AnimalVeterinary/GuidanceCompliance Enforcement/GuidanceforIndustry/ucm123631.htm. Accessed January 9, 2017.

mental decision by the appropriate regulatory authorities. The developer would be notified of the determination and provided a pointer to more information about the appropriate agency and point of contact. Consistent with the guidelines developed by the International Risk Governance Council (Renn, 2005; IRGC, 2015) and the 1996 National Research Council report *Understanding Risk* (NRC, 1996), the level of participation of outside experts, stakeholders, and interested and affected parties and the level of effort for both developer and the regulatory agencies would increase from the bin for *familiar and not complex* products through to the bin for *unfamiliar and complex* products. The model used by the International Risk Governance Council for managing different types of risk problems is illustrative of the degree of agency and stakeholder involvement that may be necessary depending on a product's familiarity and complexity (Table 5-1). Thus, as complexity increases, so does a need for engaging external experts, industry stakeholders, and interested and affected parties in the dialogue.

The method and amount of public engagement for future biotechnology products would also vary according to familiarity and complexity. Products that are *unfamiliar and complex* could require external peer review and input from interested and affected parties. The peer review and input from parties would be facilitated by one or more of the appropriate/relevant agencies—broad agency engagement is desirable if additional, future product types are envisioned to have different regulated uses. Peer review or public engagement could be designed to protect confidential business information as needed. External peer review or external party input could be used for problem formulation and then for the subsequent draft risk assessment. Iterative risk-assessment and risk-management decision making may be appropriate based on the nature and extent of the estimated risk and associated uncertainties. Peer review and engagement by external parties on potential future products could also be initiated by the regulatory agencies based on horizon scanning. Undertaking such proactive, pilot projects will increase preparedness.

A product developer would not have to use the voluntary point of entry and could independently determine whether or not their product is regulated. If it determines the product is regulated, the developer could independently ascertain the statute(s) and agency (or agencies) appropriate for the situation and directly submit the product for review; if an incorrect determination was made, the developer could subsequently work with the regulatory agencies to route the submission to the appropriate agency. A developer could use in-house expertise, private-sector consultative legal and regulatory-science expertise, or both to provide general and product-specific guidance. At the time the committee was writing its report, this practice was common within the business community for

TABLE 5-1 Escalating Levels of Expert and Stakeholder Involvement and Effort in the Management of Different Types of Risk Problems

Risk Problem	Simple	Complexity	Uncertainty	Ambiguity
Actors	Regulatory Agency Staff	Regulatory Agency Staff External Experts	Regulatory Agency Staff External Experts Industry Stakeholders Affected Parties	Regulatory Agency Staff External Experts Industry Stakeholders Interested and Affected Parties
Remedy	Statistical Risk Analysis	Probabilistic Risk Modeling	Risk Balancing with Probabilistic Risk Modeling	Risk Tradeoff Analysis and Deliberation with Risk Balancing and Probabilistic Risk Modeling

SOURCE: Adapted from Renn (2005).

dealing with regulatory issues under the Coordinated Framework. The committee recognized from the presentations it heard from startup companies and small firms and from its deliberations that this approach is currently used, especially with those businesses with some degree of in-house experience and resources, but it is not easily or routinely used by a host of smaller enterprises entering into the biotechnology product space. Therefore, a formal structure governed through collaboration among the regulatory agencies, such as that described in Box 5-2, is an important consideration for future products of biotechnology.

Risk Analysis and Public Participation

In updating the Coordinated Framework and presenting the *National Strategy*, the federal agencies have taken into account the nature of biotechnology products that were visible in 2016. In looking at the products of biotechnology that are likely to emerge in the next 5–10 years, Chapter 2 describes some of the features of future products that will challenge the system and Chapter 4 articulates some of the challenges in applying the Coordinated Framework. In moving from products that are in columns B and C to those in column D of Figure 2-6, it will be important for agencies to be prepared for products that involve substantial internal complexity, complex interactions with the environment, relatively few or no comparators to nonbiotechnology products for use in risk analysis, and have little similarity with existing biotechnology products. In this section, the committee articulates some of the features of these types of products and provides possible perspectives on how risk analysis could be performed.

Natural-science evidence, social and economic evidence, and values all influence risk analyses for future biotechnology products (NRC, 1996; Thompson, 2007; Kuzma and Besley, 2008; IRGC, 2015). Given the diversity, pervasiveness, and power of new biotechnology methods and products, public concerns that have followed and are likely to continue to follow biotechnology products into the market, and increasing complexities and uncertainties associated with anticipating the human health and environmental effects of unfamiliar and unconfined releases of biotechnology products or living genetically engineered organisms, it will become increasingly important to develop oversight systems that are adaptive, iterative (learning from past experiences or new data and information, or new concerns that emerge), and engage a wider range of expertise (Wilsdon and Willis, 2004; Stirling, 2007; Meghani and Kuzma, 2011; Ramachandran et al., 2011; Marchant and Wallach, 2015).

The social-science literature suggests several middle-ground approaches to framing future conversations that could increase public confidence in the oversight of products of biotechnology. Paradigms of responsible innovation (Stilgoe et al., 2013), critical realism (Freudenburg, 1996), strong objectivity (Harding, 1996), and analytical–deliberative risk analysis (NRC, 1996) all recognize that what is available in the empirical world is useful but also that human interpretation brings meaning to that evidence and is just as crucial. These frameworks address concerns of multiple stakeholders and disciplines, consider what evidence or risk-mitigation strategies could help address those concerns, anticipate which of biotechnology products or processes should receive greater regulatory scrutiny and which should receive less, and prepare for future concerns and products by beginning the deliberations and identifying regulatory-science needs further upstream in product development (Barben et al., 2007; Kuzma et al., 2008; Guston, 2014). Life-cycle analysis of energy, water, and chemical inputs and outputs, risk–benefit analyses, the risk of doing nothing compared to alternatives, and cultural considerations (especially to disenfranchised groups) could also be part of the oversight. These approaches will be especially important for open-release, unfamiliar applications of biotechnology such as deextinction and gene drives and for other future biotechnology products that have complex interactions and risk pathways. These approaches may also be important to provide an opportunity for future governance that is science informed, public guided, and value attentive.

A common recommendation from prior National Academies reports is the need to increase public participation in the regulatory process (see, for example, NRC, 2008). As indicated already above, it is likely that future products of biotechnology could be controversial due to their complex interactions with the environment and society, and the committee anticipates that additional concern from the public will be a common feature of many future biotechnology products. Increasing public participation in the regulatory process raises the possibility of increases in agencies' costs and inefficiencies in the overall decision-making process. Other parties may be concerned that such an approach could be fraught with complications in ensuring a balanced representation of viewpoints.

Oversight of complex and interdependent activities by their very nature requires input from multiple developers and interested and affected parties to develop and revise approaches over time. Formulating an agency approach for such complex scenarios "in secret" (bureaucratic closure) or behind closed doors with a select group of developers or interested parties (private bureaucratic learning) increases the risk of failure due to retribution from excluded participants or lack of agency capacity or statutory jurisdiction to address all the tasks needed for implementation (Moffitt, 2014). To explore these concerns, the committee considered ways additional external participation may be incorporated in those future biotechnology products that are *unfamiliar and complex*. The proposed uses of public participation and external peer review are generally consistent with a paradigm articulated by Moffitt (2014; see Figure 5-3). This paradigm acknowledges two dimensions in an agency's regulatory decision making: (a) implementation independence to implementation interdependence and (b) lack of or incomplete information and understanding to full information and understanding.

In cases where a regulatory agency has high interdependence (for example, it is supporting a future voluntary, self-regulation system where the agency depends on technology developers for oversight implementation or its decision must be integrated with input from another agency) but has a high level of information, the agency could distribute information to developers that the agency depends on as well as to the public to transparently share current information strengths and limitations to develop the oversight approach ("participatory bureaucracy"). A participatory bureaucracy can increase the chance of success by exposing any information gaps and including the values of the developers and interested and affected parties in a voluntary program. When an agency has high interdependence and a low level of information, employing participatory bureaucracy can create new information by engaging input from experts, developers, and interested and affected parties. In cases where an agency has high interdependence, a lack of knowledge, and employs a closed process for making a decision, it increases the likelihood of "eroding bureaucratic administration when it prevents bureaucrats from acquiring needed expertise, from considering helpful alternatives or from learning from experience and mistakes" (Moffitt, 2014:47). Furthermore, a bureaucratic closed or private learning approach to developing and implementing an oversight approach could increase the likelihood of challenges (legal or otherwise) due to the opaque nature of decision making and the exclusion of informed input from groups (developers and interested and affected parties alike) outside the bureaucracy or outside the limited set of groups that were invited by an agency for consultation. This perspective is consistent with the findings of numerous science and technology policy scholars who have looked at biotechnology and concluded the same (Harding, 1996; Bozeman and Saretwitz, 2005; Thompson, 2007; Meghani and Kuzma, 2011; Kuzma, 2013).

One possible approach for including external input is through the use of advisory boards implemented through the Federal Advisory Committee Act (FACA) to assess strengths and limitations of alternative risk-analysis approaches by engaging developers and interested and affected parties and gaining external scientific peer review. The employment of a FACA process does not come time or cost free, and the costs of implementing such an approach could outweigh the benefits of gaining input and advice and sharing information (Balla and Wright, 2003; Box 5-3). There are also criticisms of the use of FACA groups that include allegations of privileging specific interest groups,

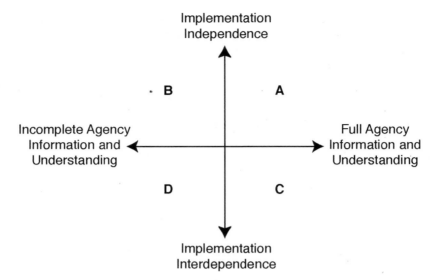

FIGURE 5-3 Participation in American bureaucracy by task-specific information and implementation conditions.

NOTES: In cases where an agency has regulatory independence and full knowledge to undertake its regulatory tasks, it could opt to make decisions "in house" and not share knowledge or decision-making logic publicly in a meaningful way (quadrant A, "bureaucratic closure"). If the agency has high independence and incomplete knowledge, it could gather information from outsiders "behind closed doors" (quadrant B, "private bureaucratic learning"). In situations where the agency has high interdependence, it could opt to share information, realizing the need to protect confidential business information, and create a process to take in and address additional information (quadrant C, "participatory bureaucratic oversight"). In cases where an agency has high interdependence but limited knowledge on an issue, it could publicly acknowledge its lack of information and initiate a public learning process to inform the future decision (quadrant D, "participatory bureaucratic learning").

SOURCE: Adapted from Moffitt (2014).

limiting use of agency expertise, promoting secrecy rather than transparency, or driving acceptance of an agency's position rather than receiving advice and input on an agency's options (Moffitt, 2014). These concerns need to be addressed in an advisory process, and the committee suggests the agencies consider the program-management and conduct-of-practices principles provided in the 2008 National Research Council report *Public Participation in Environmental Assessment and Decision Making*, which include

1. Clarity of purpose and diagnosis of context,
2. Commitment to use the process to inform their actions,
3. Adequate funding and staff,
4. Appropriate timing in relation to decisions,
5. Focus on implementation,
6. Monitoring of the process and adjusting tools and techniques as needed,
7. Inclusiveness of participation,
8. Collaborative problem formulation and process design,
9. Transparency of the process, and
10. Good-faith communication.

BOX 5-3
Studies Evaluating the Costs and Benefits of Using Federal Advisory Committees

There appear to be a small number of published studies that quantitatively compared the costs or time in making regulatory decisions with and without receiving input and advice from a federal advisory committee authorized under the Federal Advisory Committee Act (FACA). Studies that have been undertaken are limited to the FDA FACA committees that provide advice and input on new drug applications and premarket device applications (Lavertu and Weimer, 2010; Moffitt, 2014). These committees are comprised of individuals with expertise regarding the drug or device application at hand and include scientific and medical experts as well as representatives of relevant industry, consumer, or patient groups. The committees are typically used when the means to address the uncertainty in the costs and benefits of the decision exceed FDA's technical capabilities or capacity (low knowledge) and/or when the decision has numerous interdependent tasks (i.e., implementation tasks that may be required of the drug manufacturer, physicians, etc.), which if not addressed effectively increase the likelihood of implementation failure.

Moffitt (2014) reported that when taking into account drug risks, those drugs that received an advisory review were significantly less likely to have black box warnings[a] required on their labels due to post-market adverse effects and less likely to be withdrawn from the market. Although the costs of using an advisory committee were not compared to the costs of developing and issuing a black box label or implementing and enforcing a market withdrawal, it seems reasonable to assume such costs to FDA and the risks to users of the products would be higher than a scenario where the likelihood of future adverse effects were identified prior to market approval.

Lavertu and Weimer (2010) analyzed advisory committee reviews of new drugs and devices over the period 1997–2006. These authors reported that the time taken to approve a drug was not significantly longer when input from the FACA committee was requested; however, decisions for medical devices did take longer when referred to the advisory committee. For new drugs, it took FDA an average of 526 days to make an approval without advice from the committee and an average of 525 days when the committee was convened. Of note, in those instances when the advisory committee reached consensus on an approval decision, the time for an FDA decision was 456 days on average, which includes the time to manage the advisory committee process. These findings indicate that consistent external advice and input can significantly reduce decision time, but even in those cases where committee advice does not resolve the uncertainties identified by FDA, presumably the input further enhances the agency's decision documentation but not at the expense of increasing the overall time to a decision. A comparison of the time taken to make a rejection decision with and without advisory committee advice was not possible since the number of rejection decisions and the time taken for those decisions when the advisory committee was not convened are not readily available.

While there are no published analyses of the costs and benefits of EPA's use of its Scientific Advisory Panel (FIFRA required a FACA committee to provide external, scientific peer review to EPA on pesticide risk-assessment issues), decision review times are estimated to be 6 months longer with a science advisory panel (SAP) review of a plant-incorporated protectant risk assessment and registration service fees (which cover, in part, EPA's costs for reviewing pesticide applications) are approximately $60,000 higher.[b] EPA indicates that its use of the SAP is for when scientific data for a decision are complex. EPA further notes that it

> often seeks technical advice from the Scientific Advisory Panel on risks that pesticides pose to wildlife, farm workers, pesticide applicators, nontarget species, as well as insect resistance, *and novel scientific issues surrounding new technologies* (emphasis added). The scientists of the SAP neither make nor recommend policy decisions. They provide advice on the science used to make these decisions. Their advice is invaluable to the EPA as it strives to protect humans and the environment from risks posed by pesticides. Due to the time it takes to schedule and prepare for meetings with the SAP, additional time and costs are needed.[c]

---

[a]Black box warnings appear on prescription drug labels to call attention to serious or life-threatening risks. See FDA (2012).

[b]PRIA Fee Category Table – Biopesticides Division – PIP. Available at https://www.epa.gov/pria-fees/pria-fee-category-table-biopesticides-division-pip. Accessed September 14, 2016.

[c]Ibid.

Ginsberg (2015) also noted that an effective FACA process entails securing clear agency commitment; finding a balance between responsiveness to the agency and independence; leveraging resources through collaboration with similar FACA groups; and evaluating a FACA group's usefulness to identify future directions or improvements.

The committee also reviewed the potential role stakeholder rulemaking and private standard setting could play in enhancing efficiency in the proposed decision-making framework. The committee concluded in some circumstances these approaches may be preferable to a FACA process or a process in which the agency independently establishes and implements a regulatory process or requirement (see Box 5-4).

TECHNICAL TOOLBOX AND CAPABILITIES FOR
RISK ASSESSMENT AND REGULATORY SCIENCE

The committee synthesized information received during public meetings, webinars, and the results of National Academies reports (NRC, 2013; NASEM, 2016a,b), symposia (Drinkwater et al., 2014; Roberts et al., 2015), and relevant publications to identify gaps in risk-analyses tools and possible approaches that could be advanced to close these gaps. Addressing these gaps through a design-build-test-learn paradigm can help support development of a responsive research agenda and staffing plans for enhancing existing capacity, capability, and expertise needed for efficient and sound evaluations of future products of biotechnology. Separately, the tools and technologies used by product developers could be enhanced to ensure a higher probability of success in navigating the regulatory system. Finally, the committee identified specific needs in the area of regulatory science. The committee recognizes that the tools and techniques described here require a depth of data and analysis that may be inconsistent with the degree of risk that can be anticipated for many future products of biotechnology. Their blanket application would be inconsistent with tiered risk-assessment strategies and with the capacity available in the public and private sectors. The intent of the committee is to highlight the emergence of these approaches and their application for clarification of regulatory understanding of future products, especially when qualitative or deterministic risk analyses are uncertain as to whether they are incorporating well-characterized, worst-case assumptions to support the safety standard associated with risk-management decision criteria (see Box 4-2).

Implementation of Probabilistic Risk Analyses Associated
with Future Products of Biotechnology

As discussed in Chapter 4, probabilistic risk analysis has not been widely used in the regulation of biotechnology products. However, the use of quantitative risk assessment is well established in many fields and is applied to questions of ecological, food-safety, biosecurity, and biological risk. The most common application of quantitative risk assessment for biotechnology products is for the purposes of insect-resistance management of *Bt* crops (Storer, 2003), but there are also examples of quantitative nontarget-species ecological risk assessment (Sears et al., 2001), dietary exposure assessment (Exponent, 2005), and endangered-species risk assessment (Peterson et al., 2006) that have been used for regulatory decision making. The Coordinated Framework would benefit from fuller implementations of probabilistic methodologies when appropriate in light of challenges to the regulatory system that are expected to occur. This section proposes the need for more probabilistic risk assessments than were conducted when the committee was writing its report. In addition, this section discusses the need to conduct risk analyses that are proportional to the quantitative risks assessed to prevent the problem of "second-order risk" (that is, the risk of missing a significant risk versus the risk of overanalyzing a negligible risk).

BOX 5-4
Stakeholder Rulemaking: A Potential Process for Public Participation

A potential alternative or complement to using a Federal Advisory Committee Act (FACA) process or public rulemaking could be stakeholder rulemaking. As summarized by Weimer (2006) and Weimer and Wilk (2016), Congress must authorize an agency to employ stakeholder rulemaking, including use of agency funds for implementing and supporting the process to ensure continuity. Congressional authorization creates a nongovernmental organization (NGO) that is charged to develop and adopt rules for a specific function or activity. Typically, the NGO has an executive board and supporting committees that include experts in the relevant fields as well as interested and affected parties. Meetings of the NGO's committees are open to the public. The NGO has an established charter to formulate rules under a specified voting procedure. A rule developed by the NGO based on majority vote can be implemented immediately because the necessary actions required by the rule are carried out by the members of the NGO. A stakeholder rule does not impose legally binding rules. However, if members of an NGO reach a consensus on a private rule, an agency could proceed with formal rulemaking. Current examples of stakeholder rulemaking include those made by the Internet Corporation for Assigned Names and Numbers, the Organ Procurement and Transplantation Network, and eight regional fish councils that have sole jurisdiction in devising fishery regulations.

A potential strength of stakeholder rulemaking is technical efficiency. Stakeholder rulemaking is likely to be technically efficient when the major stakeholders in an NGO have a stake in the outcome and the required expertise can be employed more rapidly than possible by the agency alone or through public rulemaking or a FACA process. It is important to note that while stakeholder rulemaking can be more technically efficient, the outcome of the rule may or may not be desirable to all interested and affected parties (Weimer, 2006).

A less expansive form of stakeholder rulemaking is private standard setting (Weimer, 2006). Organizations such as UL (previously known as Underwriters Laboratories), the American Society for Testing and Materials, and the American National Standards Institute maintain a wide range of standards. Industry committees that coordinate generally recognized as safe (GRAS) analyses and propose determinations for food flavoring[a] and cosmetic ingredients[b] have been in place for decades. Private standards are not legally binding and do not involve an explicit delegation of rulemaking authority, but they can be adopted by regulatory agencies and be required in private contracts. The standards can support market claims for products, thereby providing a competitive advantage, which in turn can drive compliance.

In the context of the committee's illustrative decision-making framework for the Coordinated Framework, a stakeholder rulemaking process may have merit for classes of products that may not clearly fall under a specific statute (for example, products that are not plant pests) or for products that may potentially fall under multiple statutes. Stakeholder rulemaking could also be employed to more efficiently develop and modify decision making for classes of products within the *familiar and noncomplex* bin, to optimize notification procedures and establish protocols for data sharing among developers. Private standard setting could be employed for future consumer products and food additives, to establish testing methods for data needed to support risk assessments, and for establishing information knowledge bases and metadata requirements to support developers and agency risk assessors.

---

[a]See, for example, Flavor & Extract Manufacturers Association, About the FEMA GRAS™ Program. Available at http://www.femaflavor.org/gras. Accessed January 9, 2017.

[b]See, for example, Cosmetic Ingredient Review. Available at http://www.cir-safety.org. Accessed January 9, 2017.

Biotechnology products are diverse and therefore may vary in their associated risks. That is, some biotechnology products could be used with a lower probability of risks (for example, crops genetically engineered with insect resistance or bacteria within bioreactors that are similar to engineered products already in commerce with a familiar risk profile) while other biotechnology products may have uncertain risks at greater spatial and temporal scales to consider (for example, organisms with gene drives or genetically altered bacteria released into an open environment). For better understood products, available information from analogous systems or organisms or analyses of the published literature (for example, meta-analyses or pathway analyses) may suffice to assess associated risks. In contrast, unfamiliar products for which there is not a sufficient baseline of information may require more sophisticated quantitative analyses to estimate their associated risks. In cases of high uncertainty, data may need to be generated to be able to estimate risk with acceptable confidence.

Probabilistic approaches are summarized by recent National Academies reports and the scientific literature (for example, Suter, 2007; Warren-Hicks and Hart, 2010; NRC, 2013; NASEM, 2016a). A National Research Council report (NRC, 2013) described three principle steps in preparing a probabilistic risk assessment, which the present committee concludes are also applicable for assessing risks of biotechnology products:

- Describe uncertainty for variables with distributions (realizing all variables in a model need not require the same degree of data intensity).
- Propagate uncertainty through distributions of exposure and effects variables.
- Integrate exposure and effect estimates to calculate risk probabilities.

Example calculation methods include Monte Carlo analyses, Bayesian methods (some of which also use Monte Carlo simulations), and uncertainty bounding analyses (Warren-Hicks and Hart, 2010; NRC, 2013; NASEM, 2016a). At the time the committee was writing its report, probabilistic approaches were rarely implemented in ecological risk assessments for chemical pesticides (NRC, 2013). On the basis of the committee's limited survey of existing risk assessments, environmental assessments, and environmental impact statements for biotechnology products, probabilistic analyses have seldom been undertaken. The committee believes that the risk analyses customarily conducted in environmental assessments and environmental impact statements required by NEPA may be inadequate to characterize the risks of certain future products of biotechnology. The committee found no statutory restriction that precludes the regulatory agencies from conducting quantitative risk assessments.

The further need for quantitative approaches for human health and environmental safety involves questions of multiple exposures, complex mixtures, and vulnerable populations, which represent broad stakeholder concerns often considered to be inadequately captured in risk analyses. A recognized need in quantitative risk assessment is improved cumulative risk assessments combining risks of aggregate exposure to mixtures that include all routes, pathways, and sources (NRC, 2009). Revision and extension of existing approaches to cumulative risk assessment will be needed to fully analyze future products of biotechnology.

Even under conditions of unfamiliarity and complexity, probabilistic risk assessments can be used to identify where information is missing. Several researchers from the Commonwealth Scientific and Industrial Research Organisation in Australia have used a combination of stakeholder, expert, and public input; Bayesian elicitation; and fault-tree analysis to develop and quantify (with uncertainty) risks from genetically engineered fish (Hayes et al., 2014) and genetically engineered insect pests (Murphy et al., 2010; Murray et al., 2016). These serve as models for both probabilistic risk analysis and public engagement in an analytical–deliberative process (NRC, 1996).

The quantitative risk analyses discussed above support the means to refine risk analyses by incorporating new data through iterative assessments and enable risk assessors, risk managers, and stakeholders to refine risk-management options as needed to meet the regulatory standard for a safety finding (NRC, 2013). An established probabilistic risk-assessment framework for a given product for a suite of use-pattern scenarios (such as those proposed in Chapter 4) also can facilitate timely updates to risk estimates based on new information and help form hypotheses for causes of unexpected risks that may emerge.

The regulatory agencies vary in the degree in which risk-assessment tiers can be implemented in concert with risk-management needs. For example, EPA's pesticide risk-analysis approach uses risk-assessment tiers with increasing resolution based on the results of lower-tier risk assessments, the endpoints of concern, and the nature of requested use patterns. In addition, EPA can implement a higher-tier risk assessment to refine an existing assessment, based on adverse-effect information submitted through Section 6(a)2 of FIFRA (40 C.F.R. Part 159)[5] and Section 8(e) of the Toxic Substances Control Act (TSCA).[6] In addition, FIFRA requires EPA to reevaluate registered pesticides at least once every 15 years to ensure the existing risk analysis and regulatory decision are current with the state of the science and policy (40 C.F.R. Part 155).[7] It is more difficult for USDA–APHIS to implement iterative risk analyses because the agency as of 2016 did not have authority to reassess products once they were deregulated (McHughen and Smyth, 2008).

Ecological Risk Assessment Within the Context of Future Biotechnology Products

Ecological risk assessment for future biotechnology products and their release scenarios will necessitate more emphasis on measurement and modeling of effects to populations and communities within landscapes than has been necessary with biotechnology products regulated in the 1990s and 2000s. Further challenges arise regarding the biological responses that are used to determine effects to entities of concern for ecological risk assessment (Forbes et al., 2001). The relationship between lethal and sublethal effects to individuals and the survival and reproduction of populations is a continuing uncertainty in the ecological risk-assessment process (NRC, 2013). Typical laboratory toxicity tests focus mostly on individuals through measurements of lethality, growth rate, or both and occasionally have been extended to measures more directly representative of populations (reproductive success). Field-scale studies may more fully encompass populations and communities through consideration of abundance for greater numbers of taxa (Naranjo et al., 2005). An emphasis in ecological risk assessment on individuals in and near production fields is logical and has been successful in understanding single-stressor effects within fields of genetically engineered crops as of 2016.

The environments in which some future biotechnology products will be deployed, however, will represent a dynamic temporal–spatial mosaic where multiple novel stressors with sometimes overlapping effects are being introduced at large geographic scales such as a watershed or geopolitical region and where there may be incomplete quantitative description of effects on populations. Simple approaches for lower-tier screening that may consider effects that may scale in the environment include simple functional ecology models based on life statistics for trophic–functional types to determine the magnitude of effect necessary to become evident in the ecosystem (Raybould et al., 2011) or considerations of aggregate sensitivity to species occurring within the environment (Wolt and Peterson, 2010; Wolt, 2011). These approaches, however, still place boundaries on the system to encompass limited spatial and temporal scales, thus leaving unanswered changes occurring in the contiguous landscape over time. Future products of biotechnology designed for open release in minimally managed or unmanaged environments will introduce an increasing diversity of potential environmental stressors that will necessitate improved ecological risk assessment to forecast potential effects with a view toward understanding and managing ecological services at the landscape level. The limitations of species-specific modeling and measurement in landscapes

---

[5]Incident Reporting by Pesticide Manufacturers/Registrants. Available at https://www.epa.gov/pesticide-incidents/incident-reporting-pesticide-manufacturers-registrants. Accessed September 14, 2016.

[6]Reporting a TSCA Chemical Substance Risk Notice. Available at https://www.epa.gov/assessing-and-managing-chemicals-under-tsca/reporting-tsca-chemical-substantial-risk-notice. Accessed September 14, 2016.

[7]Registration Review Process. Available at https://www.epa.gov/pesticide-reevaluation/registration-review-process. Accessed September 14, 2016.

argue for a more generalized approach focusing on functional groups and their distribution and density among elements within the landscape (Caron-Lormier et al., 2009, 2011) for certain products intended for open release in low-management environments.

To explore how ecological risk assessment might be applied, government agencies could pilot advances in ecological risk assessments and benefit analyses for open-release products expected over the next 5–10 years. Aspects to be explored could include external, independent peer review, public participation, and whether agencies' staff will need new skills on quantitative risk-assessment practices. Risk assessors have used stakeholder and public-informed processes for broader ecological risk analyses of genetically engineered crops and fish to incorporate on-the-ground knowledge and values associated with multiple ecological and societal risk-assessment endpoints, especially in stages of problem formulation and risk-management options (Nelson et al., 2004; Kapuscinski et al., 2007). Multicriteria approaches to choose ecological indicators for risks to biodiversity and fault-tree analysis have also been applied to genetically engineered plants (Andow et al., 2013). These examples point to integration of multiple risk-assessment endpoints, modeling approaches, and societal values in the risk-assessment process. The agencies would benefit from a review of over two decades of literature on iterative and engaged methods of risk analysis for transgenic organisms.

Public–private investments in new environmental risk–benefit analytical approaches, including the identification of information needs; the development of assay methods and laboratory and field-study designs and monitoring protocols; and models (conceptual through computational) to inform risk assessments across appropriate biological, spatial, and temporal scales can also be used to address potential ecological outcomes associated with future open-release biotechnology products.

Enhancing the Capabilities, Expertise, and Tools of Regulatory Agencies

In the previous sections of this report, needed risk-analysis knowledge and technological capabilities were noted. Chapter 2 describes types of future biotechnology products, many of which will not have obvious comparators to nonbiotechnology products and in turn may require a new generation of risk-analysis approaches. Some of the use patterns for future products also highlight the need for developing spatially and temporally explicit risk-assessment capabilities. In addition, Chapter 2 points to the potential increase in the sheer number of products that may need to be assessed in the future, which highlights the need for an effective, high-throughput risk-analysis system. In Chapter 4, the need for probabilistic assessments to better interpret comparative risk assessments and management options was introduced. The above section "Consistent, Efficient, Effective Decision-Making Processes for Future Products of Biotechnology" also raises the need for assessing similarities and differences among biotechnology products and anticipates a stratified assessment process that in some cases will be highly reliant on access to existing risk-analysis data or data summaries for biotechnology products already in the market. To this end, a suite of publicly available physical and computational models and methodologies that can be accessed for risk assessments with different degrees of complexity would be helpful. Examples of sampling designs and indicators to support post-market surveillance and monitoring programs would also be beneficial.

To organize the discussion on risk-analysis tools needed for products expected in the next 5–10 years, the committee adapted categories of future research needs prepared through workshop deliberations addressing the need for a research agenda exploring the ecological implications of synthetic biology (Drinkwater et al., 2014) and a workshop and Delphi study on synthetic-biology governance (Roberts et al., 2015). The research areas identified to address gaps in risk analyses include many of those the committee sought information for in its request for information (RFI) to federal agencies (see Chapter 4 and Appendix C): comparators, off-target gene effects, and phenotypic characterizations; gene fitness, genetic stability, and horizontal gene transfer; control of organismal traits; monitoring and surveillance; modeling and life-cycle analyses; and economic and social costs and benefits. The responses to the RFI indicate that some work is being done in these areas, but the committee thinks it is likely insufficient for the number and kinds of biotechnology products the agencies can expect to see. The committee also identified molecular characterization and standardization of risk-analysis methods and data management as areas in need of research.

Comparators, Off-Target Gene Effects, and Phenotypic Characterization

As discussed earlier in this chapter, there is a need to advance quantitative comparisons that can facilitate assessment of future biotechnology products. A key characteristic of the risk-assessment process in use at the time the committee was writing its report was comparability between biotechnology products and their nonbiotechnology counterparts. However, as noted in Chapter 4, the use of nonbiotechnology comparators is becoming more challenging. Transformations can be made in host organisms that are not well characterized, and there may not be baseline data on the nontransformed counterpart host. Furthermore, some new biotechnology products may contain only synthetic DNA, which would have no nonbiotechnology counterpart. Therefore, the idea of "comparator" may need to expand to include similar existing biotechnology products with which regulatory agencies already have experience.

Methods to quantitatively compare products will be needed for determining which bin is appropriate for a new product; selecting data from other product data sets for screening-level risk assessments or problem formulation; selecting data to use in effects or exposure analysis steps in a risk assessment; and selecting data to generate a risk characterization of a new product and/or place a characterization of a new product in context with an existing, similar product. Elements of a risk assessment are typically considered against baseline nonbiotechnology comparators to address whether, other than the intended change of the modification, the observed attributes of the transformed organism represent a substantive change relative to the comparators. The degree of uncertainty in making comparisons will need to be quantified given that different risk-assessment steps or scenarios can tolerate different levels of uncertainty at key decision points; that is, findings in risk assessments are worded in language specific to the statute under which they are being evaluated, but in every instance represent an "as safe as" determination.

Research approaches need to address questions raised by risk assessors and managers concerning comparisons that are context specific and reflect the need to assess similarity across levels of biological organization and spatial and temporal scales. Issues and questions raised in risk analyses can inform development of a research agenda. For example, products may be comparable at one level of biological organization, but not at other levels (for example, target genes and off-target genes and their expression, to protein structure and function, to biochemical function, tissue/organ function, organismal, population, and community effects). There may be variability in comparability for the same biotechnology product under different environmental conditions. Products may be comparable in terms of affecting a common physiological function, but the mechanisms by which they initiate the physiological responses could be very different. Some products may be comparable in terms of the genes being manipulated, but the commercial application of the products and their use patterns may be different. Depending on the risk-analysis question, the products may be considered comparable (that is, two open-release products initiate perturbations in the same invasive weed organism in a similar manner, but with some differences in off-target gene effects), but in the context of their environmental-effects analysis and their impact on nontarget organisms they may or may not be comparable: If one product has been deployed in Gulf Coast estuaries, how comparable will its effects be to the other product's effects if it is intended for open release in estuaries along the U.S. mid-Atlantic coast?

Research in this area will also need to support computational approaches for estimating missing data from data available for existing, comparable products and identifying gaps in specific information that may require targeted testing. A systematic approach, taking advantage of horizon scanning, to establish the biological knowledge bases needed to inform computational similarity analyses systems and develop decision-support systems to facilitate analyses will also be needed.

In addition to comparators, research on phenotypic characterization is also needed to advance understanding of trait function and potential ecological consequences over the short and long term as well as understanding on how environmental context can affect phenotypic expression.

Gene Fitness, Genetic Stability, and Horizontal Gene Transfer

Engineered organisms that reproduce can suffer mutations that affect the physiology of the organism, leading to the potential for "instability" in the genome (engineered genetic constructs mutating in ways that could cause loss of function). In addition, many organisms can incorporate DNA from their environment, leading to the possibility of horizontal gene transfer.[8] Techniques to measure these properties, including how these properties may vary with different environmental interactions, are needed. This research area includes evaluation and advancement of environmental models to assess properties; engineering for unanticipated interactions; developing standardized metrics and quantitative thresholds; and the interplay of fitness and stability, especially if an organism loses its containment mechanism.

Future approaches for risk assessment can be more streamlined, less costly, more comprehensive, and unbiased by utilizing state-of-the-art assay tools—for example, automated high-throughput biochemical assays, next-generation DNA sequencing, and advanced mass spectrometry technologies—integrated with high-capacity data storage and analytics. Rather than obtaining a targeted snapshot of single-few genomic loci via Southern blot, polymerase chain reaction, or Sanger sequencing to characterize a genetic modification, risk assessment at the molecular level should leverage recent advances in next-generation DNA sequencing and associated whole-genome sequence information to obtain an unbiased assessment of both the targeted and off-target genetic modifications in the species, whether altered by biotechnology or not (Pauwels et al., 2015). Similarly, untargeted mass spectrometry for metabolomics and proteomics is one approach for enhanced safety assessment of biotechnology products because it provides an unbiased assessment into potential pleiotropic effects derived from a modified organism (Ryals, 2016).

Control of Organismal Traits (Containment and Confinement)

Given that open-release products are deployed in dynamic environments, quantitative assessments of the safety, security, and stability of biotechnology-derived organisms should be tailored for the proper context. Metrics to test for biotechnology-derived organisms in open environments should measure

- Intrinsic biocontainment (i.e., escape frequencies into the natural environment),
- Genetic isolation (i.e., flow of horizontal gene transfer),
- Watermarking (i.e., unique sequence identifiers in the genomes of biotechnology-derived organisms), and
- Functional impact on the environment, including on nontarget organisms.

---

[8]Horizontal gene transfer is common in nonbiotechnology organisms.

Possible areas of research include biocontainment schemes that can be adaptive to different intended applications, environmental settings, or both; establishing redundant, stacked containment approaches; and assessing the reliability of engineered reversibility.

Monitoring and Surveillance

Following completion of a premarket risk assessment, with a decision to allow the use of the products under specified conditions, there may be a need for monitoring or surveillance to evaluate specific assumptions in risk assessments, to address uncertainties in the evaluation of a risk hypothesis in an assessment, to assess the effectiveness of any required risk-mitigation measures, or all the above. In instances where products enter the marketplace through a notice to the appropriate regulatory agency, post-market monitoring or surveillance may be used to determine if future risk analyses and potential risk mitigation may be needed following use of the products (for example, cosmetics). To ensure data obtained from monitoring or surveillance address risk-management needs, designs and indicators need to be developed to directly address specific areas of uncertainties and risk hypotheses.

Examples of questions that may need to be answered through monitoring and surveillance for different types of products include the following:

1. What is the current baseline of allergic responses to certain classes of cosmetics and the contribution of specific products? Has the introduction of a new class of living cosmetics increased or decreased the rate of allergic responses?
2. Has the removal of contaminants by a consortium of microbes at a site met the remediation goals, has the consortium been confined or contained as planned, and is the habitat responding as predicted?
3. Has the introduction of a gene drive to suppress a pest population achieved the suppression as predicted, and has the ecosystem responded as predicted with the removal/suppression of the target pest? If not, why not? Has the gene drive appeared somewhere it was not supposed to (that is, in a nontarget organism)? Has the gene drive mutated?
4. Are the discharges of living engineered microorganisms into publicly owned treatment works or receiving waterbodies altering existing microbial communities in an unanticipated manner?

The sampling designs for monitoring and surveillance—for example, stratified, probability-based survey designs or fixed-site sampling programs at the national, regional, state, or watershed scale—will need to be established to address specific questions that arise for specific products or types of products. Frequency of sampling also needs to be established for addressing specific questions. Perhaps some questions concerning future biotechnology products could be integrated within existing monitoring programs (with inclusion of new indicators), while other questions may require unique monitoring programs, sampling designs, and diagnostic indicators. Although open-release products will offer little opportunity for environmental recall should unanticipated ecological effects be observed, the committee observed that environmental release in managed versus unmanaged or low-management conditions presents differences in complexity that will influence monitoring designs and potential variations in risk management.

Research to support monitoring and surveillance will be needed to assess movement and effects of specific product applications and may also be needed to provide a broad-based assessment of environmental conditions. Indicators will be needed to assess status and trends at the molecular level (for example, metagenomics) and to track changes in structural or functional attributes of ecosystems. Monitoring designs and protocols will likely be established and directed for specific issues, but research is needed to ascertain the extent to which data sets derived with different survey features and indicators can be integrated to maximize the use of available resources.

Modeling and Life-Cycle Analyses

Both physical and computational models will be needed to help inform uncertainties in risk assessments. Physical models, such as mesocosms or controlled field studies, can provide information in specific places and time periods. For example, mesocosm experiments with GE versus wild-type Japanese medaka (*Oryzias latipes*) were used to assess gene flow over time in the life cycle of the fish (Pennington et al., 2010), indicating that such studies are possible and exist in the academic literature but are not routinely used in USDA or FDA assessments with live organisms. Computational models can be used to support the development of conceptual models within the problem-formulation phase of a risk assessment and to predict ecological and evolutionary responses in other places and time frames (that is, over decades rather than several years) that cannot be evaluated with a physical model. In some cases, the findings of a computational model may be needed prior to undertaking outdoor studies to help ascertain if there is an acceptable level of risk to undertake a study. Results from a computational model can provide insights for designing experiments with physical models. Optimally, collection of data through the use of physical models and computational models develops iteratively, each informing the other (NRC, 2007). Furthermore, a 2007 National Research Council report (2007:102–103) recommended

> Using adaptive strategies to coordinate data collection and modeling should be a priority of decision makers and those responsible for regulatory model development and application. The interdependence of measurements and modeling needs to be fully considered as early as the conceptual model development phase. Developing adaptive strategies will benefit from the contributions of modelers, measurement experts, decision makers, and resource managers.

Research is needed to advance physical (for example, microcosms, mesocosms, and controlled field studies) and computational models to improve understanding of the ecological implications of genome engineering and to reduce uncertainties in predicting future ecoevolutionary dynamics over time frames of years to decades, which will support life-cycle analyses. Identifying gaps in current physical and computational models is needed to prioritize desired, future capabilities.

Physical Models. Microcosms, mesocosms, and controlled field studies are approaches to generate data that can reduce uncertainty in assessing potential effects across levels of biological organization, space, and time (Drinkwater et al., 2014). The 2016 National Academies report on gene drives articulated a phased testing approach for gene drives that includes research preparation, laboratory research, field research, staged environmental release, and post-release surveillance to gather information to support risk assessments and risk-mitigation measures to reduce potential nontarget effects (NASEM, 2016a). That report also provided examples of field and environmental field research for biocontrol and existing engineered organisms. External peer reviews of effects of herbicides in aquatic ecosystems (EPA, 2012b) and effects of insecticides on honey bees (EPA, 2012a) also provide insights on the design and execution of mesocosm and field studies that are intended to support ecological risk assessments. Clarity in ecosystem definition and model system design (for example, its size and composition), type of risk-assessment endpoints and responses measured (including recovery of community structure and function), and approaches to interpret and extrapolate data are important features of successful studies.

Although the need for undertaking field studies to evaluate future biotechnology products is recognized, EPA in 1992 determined that field studies or mesocosm experiments for pesticide registrations would no longer be required due to uncertainty in data interpretation and a conclusion that the information gained from such studies did not alter risk-assessment conclusions based on data derived from laboratory studies (EPA, 2004). The agency can, however, conditionally require mesocosm and field studies for chemical pesticides[9] and plant-incorporated protectants (Rose, 2007) on a case-by-case basis. USDA–APHIS uses a similar rationale for tiered testing regimes extending from the laboratory to the field. The current, limited experience in using results from physical models to inform ecological risk assessments indicates proactive research is needed on the development of study designs and risk-analysis methods for future open-release biotechnology products. Pilot efforts could be undertaken to develop and evaluate new approaches for using physical models to assess population, community, and ecosystem effects. Advances should be linked to the design-build-test-learn cycle and the scaled release of biotechnology products (from laboratory scale to small field trials, larger field trials, and eventually full-scale deployment). Consistent with this perspective, the 2016 National Academies report on gene drives (NASEM, 2016a) noted that support mechanisms for risk assessment, public engagement, and governance will be needed throughout a phased testing scheme.

Computational Models. Some future biotechnology products could be assessed with a high degree of specificity concerning spatial and temporal dimensions (for example, bacteria within bioreactors) while assessments for other biotechnology products have a more complex dimensionality to consider (for example, open-release organisms with gene drives and genetically altered bacteria consortia for open release). For new products, probabilistic risk assessments that can use information and methods available for analogous systems and organisms may be completed with a lower level of effort as compared to assessments involving unfamiliar products or products with more complex spatial and temporal use patterns. As the complexity of an assessment increases (dimensionality and number and nature of the risk-assessment endpoints), computational models to support more sophisticated quantitative analyses to estimate ecological risks and evolutionary responses will likely be required because existing assessments that provide baseline information or methods will be limited. Modeling will also support life-cycle analyses for existing and future products and can be used to help inform the socioeconomic tradeoffs associated with oversight decisions. These modeling efforts will entail integration with existing approaches to assess water and fossil-fuel utilization and other ecological goods and services.

The risk estimates and descriptions in human health and ecological risk assessments for existing biotechnology products are typically qualitative in nature; however, certain portions of an assessment may be quantitative, such as for estimates of human dietary exposure assessment or determining nontarget species sensitivity. The current assessments may provide a limited discussion of the uncertainties associated with risk estimates with the overall risk-assessment conclusion based on the perspective that assumptions used in a risk assessment will provide an adequate margin of safety. The influence these assumptions have on a quantitative estimate of risks needs clarification.

In the development cycle of future models to estimate risks of biotechnology products, the committee supports the 2007 National Research Council report (NRC, 2007:161) recommendation that model *evaluation*, rather than *validation*, be employed:

> Model evaluation is the process of deciding whether and when a model is suitable for its intended purpose. This process is not a strict verification procedure but is one that builds confidence in model applications and increases the understanding of model strengths and limitations. Model evaluation is a multifaceted activity involving peer review, corroboration of results with data and other information, quality assurance and quality control checks, uncertainty and sensitivity analyses, and other activities.

---

[9]See 40 C.F.R. Part 158.

Economic and Social Costs and Benefits

As discussed in Chapter 2, biotechnology products can have economic and social benefits, but they also frequently involve economic and social risks and tradeoffs. How important concerns about future biotechnology products are in comparison to the benefits provided depends on the social and cultural position of different communities, interpretation of evidence, context, and an individual's and social group's perception of risk and technologies. Research that teases out the social and economic tradeoffs involved in developing (or not developing) a biotechnology product is important for responsible decision making about technological development. However, the committee understands that social and economic research is not within the remit of every regulatory agency. Analyses that go beyond the direct health and environmental effects of biotechnology may be conducted by product developers, academe, and think-tanks. These analyses can be helpful to regulatory agencies when communicating about the possible risks and benefits involved in biotechnology products and in increasing public understanding about the science of risk assessment and the limitations of regulatory risk assessments. More research on how to consider the multiple socioeconomic, cultural, and indirect health effects of biotechnology products is needed, as these studies are not typically funded by current government programs (see Chapter 4).

Molecular Characterization as a Preliminary Assessment Tool

Molecular characterization of biotechnology products can provide important precursor information that can guide the direction and extent of human health and ecological risk assessments necessary for regulatory decisions (Corrigan-Curay et al., 2015). For instance, for the case of genome-edited plants, the use of whole-genome sequencing and/or evaluated bioinformatics models can establish the frequency of off-target mutations within the genome resulting from CRISPR-Cas9 genome editing and therefore addresses the probability for indirect downstream effects from the genome-edited product (Wolt et al., 2016). Establishing that off-target gene mutation frequencies are at or below natural mutation frequencies also indicates that nontransformed plant varieties may be appropriate comparators for genome-edited varieties. Similarly, molecular characterization can determine if CRISPR-Cas9 reagents are removed in breeding selection by establishing that transgenic elements are absent, and this can provide assurance that a gene drive has not been accidentally released as an unintended residual effect of genome engineering (Akbari et al., 2015).

More generally, advanced molecular approaches provide a possible avenue to address potential ecological risks through proper design and interdiction or elimination of poor design in biotechnology products. Such molecular characterization is critical for early screening to triage (potential) products into bins based on familiarity and/or complexity and therefore appropriately direct regulatory-science resources; this is particularly valuable as the pace of product innovation increases and stresses the regulatory system.

Standardization of Methods and Data

New approaches to conduct risk assessment will leverage state-of-the-art tools and capabilities from high-throughput and automated experimentation in genomics, metabolomics, and proteomics to site-specific and potentially national-scale monitoring programs. While cognizant of the need to establish the performance of new assay methods, the committee encourages a process for evaluating assays by determining if they are fit for their intended purpose and avoid costly and timely assay validation processes. In this regard, the committee encourages implementing an approach to establish assay performance criteria, as was being developed to evaluate bench-level and high-throughput in vitro assays for chemical risk assessments (OECD, 2014). The use of private standard setting could provide the means to increase the efficiency of establishing performance-based assays.

Comprehensive assessment of future biotechnology products will likely generate large data sets of unprecedented size and complexity that will require state-of-the-art data storage and analytics. There will be a need to enhance existing data storage and information-technology analytical capabilities to rapidly accommodate and analyze the large data sets generated from -omics approaches to assessment. There will also be a need to establish standards under which some data sets can be made publicly available, while protecting confidential information as appropriate under federal statutes. A 2009 National Research Council report concluded with respect to advanced risk-assessment methods that "there is a need for simplified risk-assessment tools (such as databases, software packages, and other modeling resources) that would allow screening-level risk assessments and could allow communities and stakeholders to conduct assessments and thus increase stakeholder participation" (NRC, 2009:10).

In response to regulatory concerns regarding the validation and integrity of proprietary data sources used for industry data analysis and risk assessment, some shared, transparent, and publicly available resources already have been developed. Three examples are

- Allergen Online, a peer-reviewed allergen list and sequence-searchable database intended for the identification of proteins that may present a potential risk of allergenic cross-reactivity curated by the University of Nebraska.[10]
- The International Life Sciences Institute crop composition database, which summarizes ranges in nutrient, toxicant, and antinutrient content of crops for use in substantial-equivalence comparisons.[11]
- The CRISPR Genome Analysis Tool curated by Iowa State University and used for design and analysis of guide RNA for minimization of off-target genome edits.[12]

Given the large amounts of data that will be generated to support modeling and monitoring efforts, some degree of standardized methodologies and information systems will be required. Issues that will require attention include standardizing notation; standardizing testing procedures and assessment paradigms; characterizing the potential impacts of similar testing protocols on risk assessments; and approaches for collecting and integrating data from existing and future risk assessments and environmental impact statements, without compromising product developers' data compensation rights when specified under a relevant statute. Another possible approach that would be enabled by common standards for data would be the creation of scientifically based, evidence-oriented "dossier" approaches for the submission of one, common scientific information and test data kit that can be used by all agencies for different purposes with different risk standards. To improve the consistency and predictability of risk assessment for future products of biotechnology, common standards for the information that is provided for different product classes as part of the assessment process could be used.

Enhancing the Capabilities and Tools of Product Developers to Enable Future Biotechnology Products to Traverse the Regulatory Path

In addition to needs in regulatory science for the regulatory agencies, there are also risk-analysis considerations and data generation that developers might employ in their design work

---

[10]AllergenOnline. Available at http://www.allergenonline.org. Accessed January 15, 2017.

[11]The International Life Sciences Institute Crop Composition Database, Version 6. Available at https://www.cropcomposition.org/query/index.html. Accessed January 15, 2017.

[12]CRISPR Genome Analysis Tool. Available at http://cbc.gdcb.iastate.edu/cgat. Accessed January 15, 2017.

to optimize efficient risk analyses when a product is submitted for regulatory review. This section discusses the apparent discontinuity between basic bioscience activities and the regulatory process for biotechnology products. It then explores the development of tools that bridge the gap between fundamental biotechnology design-build-test-learn activities and the action of performing a regulatory assessment on a specific product submission. In considering options to employ approaches described below, beginning with simpler products (such as those intended to be contained and that have only one or a few deleted genes) would support developing design-build-test-learn cycles that eventually could be scaled to the potential open release into the environment of more complicated biotechnology products.

The tools used for regulatory assessments are aligned to guidance or statutes that are often not transparent to early-stage researchers or product developers. As the scale and complexity of the development process increases, failure to incorporate elements into early-stage product candidates leads to rework, delays, or abandonment of the product candidate in the regulatory process. Alternatively, being aware of criteria considered key to assessments of safe use might allow the developer to incorporate these early in a way that facilitates safety by design. Tools that bridge early research demands with anticipation of downstream regulatory requirements can increase the efficiency, predictability, and outcomes of regulatory assessments. Several examples of horizon scanning and anticipatory governance (Guston, 2014) for synthetic-biology products already exist; for example, a policy Delphi study focused on four cases of synthetic biology to outline research needs and governance issues for each using surveys, interviews, and a workshop. At the workshop, the most important research needs and governance opportunities and challenges were assessed for biomining, deextinction, Cyberplasm, and nitrogen-fixing microbes in the presence of a group of multidisciplinary scholars and practitioners coming from nongovernmental organizations, industry, academe, and government (Roberts et al., 2015). Another example of multiparty input in horizon scanning of potential future products which considered future regulatory needs was the Woodrow Wilson Center report *Creating a Research Agenda for the Ecological Implications of Synthetic Biology*, which identified several priority research areas (Drinkwater et al., 2014). Such workshops could serve as a model for identifying the risk-assessment tools needed in early-stage research to anticipate the downstream regulatory requirements of future biotechnology products.

A key aspect in considering these tools versus those discussed in the section "Enhancing the Capabilities, Expertise, and Tools of Regulatory Agencies" are that these are intended to be anticipatory to risk assessment. They are also tied to the technical drivers (see Chapter 2) that are enabling creation of future products. The tools here are envisioned to bridge conceptual gaps that could arise when products that fall within columns C and D of Figure 2-6 are actually placed into a regulatory framework that is underpinned by practices which might not be scalable.

The first category of tools would facilitate adoption and assessment of future products destined for open release into the environment. The key gap is biology knowledge on open-release use-pattern scenarios. The tools needed would establish systematic frameworks to enable evaluation of design or deployment concepts that have been recommended for genetic biocontainment. Such examples include modern kill-switch implementation and nutritional or genetic orthogonality. These design components need to be validated in a product-dependent framework at the point of entry to the regulatory system, but more generally they need validation as types of technologies in increasingly relevant systems. These systems might include at-scale fermentation or simulated environmental releases, up to assessment in actual limited environmental release.

Likewise, creation of proven models for microbial gene flow relevant to biotechnology products in environmental scenarios is a general need that anticipates future microbial products intended for environmental release. Establishing risk-assessment frameworks and metrics involves integrating current concepts of microbial ecology and would result in qualitative or quantitative scenarios that are important to new types of products. As a given product approaches the regulatory framework, these models will provide insight to help ascertain the degree of oversight proportional to the risks posed by the specific product.

The tools of computational biology need to be focused on resolving questions and establishing evaluation frameworks directly relevant to increasing the probability of success in the regulatory framework. Currently such tools are heavily deployed on enabling early-stage discovery or development. The adaptation of these tools to the task of predictive modeling on environmental open-release scenarios would benefit later-stage risk assessment but can inform release scenario design, and possibly point to new opportunities to design, monitor, or enhance features of future products destined for deliberate release. The development of better in silico modeling systems for outcomes relevant to health, environment, and safety questions is important as the scale or complexity of systems advances. Access to such evaluated tools would enable developers to make better design decisions earlier in development and could also be designed to be responsive to advances in knowledge within the system's technical domain, such as environment or health. This could enable the developer to prescreen or iterate designs to optimize desired outcomes, which is consistent with previous calls to design safety into biotechnology products as a first priority (Kapuscinski et al., 2003).

One example that illustrates a near-term opportunity is in computational assessment of allergen potential of a gene or gene product. Some current risk-assessment frameworks require detection and assessment of "stop to stop" codon hypothetical reading frames in all six frames (Young et al., 2012), some with no minimum length (EFSA, 2011b). Any hypothetical reading frame thus detected is taken through the computational analysis for allergenicity (Schein et al., 2007; Goodman et al., 2016). The increased use of computer-aided design systems for rapid design and assembly iteration represents an opportunity to incorporate a computational assessment early in the development process to minimize or eliminate the number of these hypothetical open reading frames during initial construct design in anticipation of a future regulatory assessment process (Galdzicki et al., 2014; Christen et al., 2015; Nielsen et al., 2016). Furthermore, by incorporating such a search and creating open-source tools to eliminate undesirable features in fundamental DNA design–build software, it may enable diffusion of this aspect of safety into the community of developers regardless of their size or knowledge of the downstream regulatory assessment framework (though with the caveat that if the downstream framework is based on faulty scientific assumptions or extra factors of conservatism, designers may limit their choices for product development by accepting rather than challenging the science). Likewise, in scenarios where a community of researchers has found value in creating libraries of standardized parts or standardized parts sequences, frameworks that allow routine screening of these resources, which are themselves a product of biotechnology, increase their utility in development work and enable safety by design at the earliest stages of a project.

A particular detail with regard to the intersection of gene sequence composition and regulatory assessment could impact the deployment of safeguarding concepts such as watermarking or DNA barcodes (Gibson et al., 2010; Liss et al., 2012; Iftikhar et al., 2015). As just mentioned, sequences subject to a regulatory review process are analyzed for many features, and the use of watermarking technology could introduce features that negatively affect the regulator's analysis of the introduced genetic elements. Therefore, automated DNA design systems that could ensure optimum balance between the objective of sequence tagging and minimizing sequences of concern would be beneficial. Ease of access to such design tools which also incorporate sequence analysis schemes important to regulatory assessment could also facilitate the incorporation of such safeguarding elements by developers.

Another example of an anticipatory computational tool would be implementation of a computational framework by which novel chassis would receive a systematic taxonomic classification. The Coordinated Framework invokes genus-level classification of the host and donor in many regulatory submission documents. To the extent that such foundational information is considered essential for proper risk framing in the future, one can anticipate that developers of future products of biotechnology, which rely on novel or orthogonal chassis, will need a scientifically sound route for establishing taxonomy.

Lastly, future products of biotechnology, particularly those in columns C and D of Figure 2-6, could require development of frameworks for rationalizing use of -omics data in anticipation of their increasing use in making risk-analysis decisions. In particular, developing nonarbitrary decision frameworks for choice of comparators will have an important impact on future risk analysis as the regulatory system is faced with increasing types of hosts and increasingly novel, engineered hosts; this will be a large undertaking and is best achieved through development of risk-analysis guidance that utilizes far-reaching engagement from an array of experts with public input. Analytical and information-technology resources that support appropriate experimental designs for using -omics data in comparative work should be enabled, including development of guidance on what is important to measure by when within a development pathway—that is, what information from the suite of -omics tools would be helpful in a consultation phase prior to submission and what information will likely be required during the submission would need to be clarified.

SUMMARY AND CONCLUSIONS

As technologies and basic knowledge advance, the regulatory system needs to be able to adapt to new risks of future biotechnology products and also to adjust to well-established categories of products as their level and types of risk become better understood. A regulatory system with a greater emphasis on stratified approaches that prioritize the regulatory agencies' familiarity with a product, the complexity of the risk assessment for the product, and the anticipated risk associated with the product (that is, proportionate oversight) could contribute to meeting the increased demands on the system.

Conclusion 5-1: It would be beneficial to develop clear points of entry for biotechnology stakeholders that provide guidance, support, and direction to future product developers on the appropriate regulatory path for products of biotechnology on the basis of organism, product attributes, and release environment.

Given the diverse set of new actors who are likely to develop new products of biotechnology, it is important that there be a consistent approach to regulatory oversight that supports a product-based, science-driven risk assessment of consumer safety and environmental protection. Clear points of entry for biotechnology stakeholders might include a federally operated Web portal, interagency coordinating office, or targeted outreach efforts (for instance, to small business using the Small Business Administration networks). Stakeholders of interest include large industry, small- and medium-sized enterprises, the do-it-yourself biology community, direct-to-consumer entrepreneurs, nongovernmental organizations with interests in one or more classes of biotechnology products, and the public at large.

The development, use, and regular updating of guidance documents have proven effective and useful by EPA, FDA, and USDA in providing predictable pathways to market and increasing regulatory input quality. Future guidance documents will need to provide clear indications of the criteria that will be used to perform risk assessments and what processes and timelines will apply. Several such efforts were already under way at the time the committee was writing its report and were described in the update to the Coordinated Framework. Documents that bridge the different agencies and provide a more concise and unified description of the regulatory routes would be particularly helpful.

Conclusion 5-2: To be prepared for the anticipated profusion of future biotechnology products, the regulatory system should use scalable and proportional methods of risk assessment, capable of handling significant increases in the rate of biotechnology product innovation, the number of biotechnology products, the complexity of interactions, and the diversity of actors (who may have varying experience with the regulatory process).

On the basis of the information gathered as part of this study, the committee concluded that there is a strong possibility that the number of products per year that require federal oversight will increase and the complexity of future assessments for these products—and the associated level of effort required on the part of appropriate regulatory authorities—will also increase. Some future products of biotechnology will be familiar and fit into product categories for which there is already substantial experience and risk-analysis approaches are well defined and well understood. An approach that focuses the most attention and resources on developing risk-analysis methods for products that are unfamiliar and more complex in terms of risk analyses should be used.

Conclusion 5-3: Participatory governance processes are available for unfamiliar and more complex products of biotechnology, especially open-release products, to enhance input from experts, developers, and interested and affected parties early in the decision-making process.

Future biotechnology products and their use patterns will be increasingly dissimilar to existing biotechnology products and relatively well-understood applications; this is especially true for open-release products that may interact with the natural environment in increasingly complex ways. Participatory governance can help inform the development and implementation of an efficient process for identifying regulatory routes. To this end, approaches to efficiently address product-deployment cycles can engage diverse stakeholders and social and natural scientists with diverse expertise to establish a rigorous system based on research in social and natural sciences and practical policy experiences. Risk analyses for unfamiliar and more complex products will benefit from participatory governance by gaining a more complete appreciation of societal values to inform definition of risk-assessment endpoints in problem formulation, consideration of uncertainties in risk characterization, and formulation of risk-management options. Participation or peer review by independent experts will enable the strongest possible scientific judgments in performing risk analyses.

This conclusion is supported by a recommendation in the National Academies report on gene drives (NASEM, 2016a:10), which stated:

> Governing authorities, including research institutions, funders, and regulators, should develop and maintain clear policies and mechanisms for how public engagement will factor into research, ecological risk assessments, and public policy decisions about gene drives. Defined mechanisms and avenues for such engagement should be built into the risk assessment and decision-making processes from the beginning.

Conclusion 5-4: Ecological risk assessment provides a methodology for more quantitative risk assessments for future biotechnology products and their release scenarios but will require more emphasis on measurement and modeling of effects on populations, communities, and ecosystems.

Comprehensive, efficient, and unbiased risk analysis requires regulatory expertise commensurate with the scale and complexity of future biotechnology products. Tools that can bridge early research demands with anticipation of downstream regulatory requirements can increase the efficiency and predictability of outcomes in regulatory assessments.

Future products of biotechnology provide many opportunities for improving risk analyses. Deficiencies in risk analyses include inadequate use of planning and problem-formulation steps in risk assessments and a resulting lack of clarity in the factors considered in selecting risk-assessment endpoints and in establishing conceptual models (that is, the rationale for determining what needs to be protected from potential harm by biotechnology products; what is the object of the protection; and what is the assumed path to harm). Increasing the quantitative nature of risk analyses required by the Coordinated Framework can, along with methods to elicit probabilities and uncertainties in the absence of empirical data (for example, Bayesian methods or fault-tree assessments), contribute to utilizing proportional efforts in risk analyses.

Conclusion 5-5: There are many opportunities for enhancing the capabilities, expertise, and tools available to regulatory agencies in areas that are likely to see increased emphasis and complexity in future products of biotechnology.

Given the nature of future biotechnology products, there are a diversity of knowledge and technological gaps in current risk-analysis approaches that if addressed on a case-by-case basis could overwhelm the capacity and capability of regulatory agencies to make efficient and sound evaluations. Risk analysis must be capable of adapting and responding to the rapid pace of technology and information. The regulatory agencies may wish to consider establishing a common risk-assessment infrastructure focused on the assessment of products designed for open release into the environment. There are unique research and risk-analysis needs for future biotechnology products, but some of the needs are similar, if not identical, to needs also faced for assessing the probability of adverse effects from other nonbiotechnology stressors. Resources in national and international programs managing these efforts could be leveraged. In addition, opportunities to establish public–private partnerships to address research needs should be explored in an open, transparent process.

Future biotechnology products will be more complex in terms of their internal and external interactions, and it is critical that the agencies involved in regulation of biotechnology develop and maintain scientific capabilities, tools, and expertise in relevant areas. Furthermore, it will be essential that the agencies stay appraised of technology trends so that they can engage in meaningful discussions with technology and product developers early in the product-development cycle, where there is often the best opportunity to affect future technologies. Determination of the key areas of scientific capability will need to adapt to the emerging technologies that underlie future products of biotechnology. On the basis of the current level of federal investments (see Chapter 4), some of the key areas of regulatory-science research for the products of biotechnology likely in the next 5–10 years include comparators, off-target gene effects, and phenotypic characterization; gene fitness, genetic stability, and horizontal gene transfer; impacts on nontarget organisms; control of organismal traits; modeling (including risk-analysis approaches under uncertainty) and life-cycle analyses; monitoring and surveillance; and economic and social costs and benefits.

Conclusion 5-6: There are many opportunities for enhancing the capabilities and tools of technology and product developers to enable future products to traverse the regulatory path.

There are substantial opportunities for the use of improved methods for scientific evaluation, risk assessment, and community engagement related to future products of biotechnology that can be applied by technology and product developers. In order to ensure that the regulatory framework is able to make use of the best available tools in performing its oversight and regulation responsibilities, it will be important to invest in those tools and make them available to regulators and product developers. Areas for consideration include stochastic methods, advances in uncertainty analysis, better ways to integrate and interpret both qualitative and quantitative data, and communication strategies.

REFERENCES

Akbari, B.O.S., H.J. Bellen, E. Bier, S.L. Bullock, A. Burt, G.M. Church, K.R. Cook, P. Duchek, O.R. Edwards, K.M. Esvelt, V.M. Gantz, K.G. Golic, S.J. Gratz, M.M. Harrison, K.R. Hayes, A.A. James, T.C. Kaufman, J. Knoblich, H.S. Malik, K.A. Matthews, K.M. O'Connor-Giles, A.L. Parks, N. Perrimon, F. Port, S. Russell, R. Ueda, and J. Wildonger. 2015. Safeguarding gene drive experiments in the laboratory. Science 349(6251):927–929.

Andow, D., G.L. Lövei, S. Arpaia, L. Wilson, E.M. Fontes, A. Hilbeck, A. Lang, N. Van Tuat, C. Pires, E. Sujii, C. Zwahlen, A.N. Birch, D.M. Capalbo, K. Prescott, C. Omoto, and A.R. Zeilinger. 2013. An ecologically-based method for selecting ecological indicators for assessing risks to biological diversity from genetically-engineered plants. Journal of Biosafety 22(3):141–156.

Balla, S.J., and J.R. Wright. 2003. Consensual rule making and the time it takes to develop rules. Pp. 187–206 in Politics, Policy, and Organizations: Frontiers in the Scientific Study of Bureaucracy, G.A. Krause and K.J. Meier, eds. Ann Arbor: University of Michigan Press.

Barben, D., E. Fisher, C. Selin, and D.H. Guston. 2007. Anticipatory governance of nanotechnology: Foresight, engagement, and integration. Pp. 979–1000 in The Handbook of Science and Technology Studies, 3rd Ed., E. Hackett, O. Amsterdamska, M.E. Lynch, and J. Wajcman, eds. Cambridge, MA: MIT Press. Available at https://www.hks.harvard.edu/sdn/articles/files/Barben-STS_Handbook-Anticipatory_Governance_Nanotechnology-08.pdf. Accessed October 28, 2016.

Benedict, M.Q., and A.S. Robinson. 2003. The first releases of transgenic mosquitoes: An argument for the sterile insect technique. Trends in Parasitology 19(8):349–355.

Biello, D. August 29, 2014. Ancient DNA could return passenger pigeons to the sky. Scientific American. Available at https://www.scientificamerican.com/article/ancient-dna-could-return-passenger-pigeons-to-the-sky. Accessed January 8, 2017.

Bozeman, B., and D. Sarewitz. 2005. Public values and public failure in US science policy. Science and Public Policy 32(2):119–136.

Callaway, E. 2015. Mammoth genomes hold recipe for Arctic elephants. Nature 521(7550):18–19.

Caron-Lormier, G., D.A. Bohan, C. Hawes, A. Raybould, A.J. Haughton, and R.W. Humphry. 2009. How might we model an ecosystem? Ecological Modelling 220(17):1935–1949.

Caron-Lormier, G., D.A. Bohan, R. Dye, C. Hawes, R.W. Humphry, and A. Raybould. 2011. Modelling one ecosystem: The example of agro-ecosystems. Ecological Modelling 222(5):1163–1173.

Christen, M., S. Deutsch, and B. Christen. 2015. Genome calligrapher: A Web tool for refactoring bacterial genome sequences for de novo DNA synthesis. ACS Synthetic Biology 4(8):927–934.

Corrigan-Curay, J., M. O'Reilly, D.B. Kohn, P.M. Cannon, G. Bao, F.D. Bushman, D. Carroll, T. Cathomen, J.K. Joung, D. Roth, M. Sadelain, A.M. Scharenberg, C. von Kalle, F. Zhang, R. Jambou, E. Rosenthal, M. Hassani, A. Singh, and M.H. Porteus. 2015. Genome editing technologies: Defining a path to clinic. Molecular Therapy 23(5):796–806.

Drinkwater, K., T. Kuiken, S. Lightfoot, J. McNamara, and K. Oye. 2014. Creating a Research Agenda for the Ecological Implications of Synthetic Biology. MIT Center for International Studies, Cambridge, MA, and Woodrow Wilson International Center for Scholars, Washington, DC. Available at https://www.wilsoncenter.org/sites/default/files/SYNBIO_create%20an%20agenda_v4.pdf. Accessed August 10, 2016.

EFSA (European Food Safety Authority). 2010. Guidance on the environmental risk assessment of genetically modified plants. EFSA Journal 8(5):1879–1989.

EFSA. 2011a. Guidance on conducting repeated-dose 90-day oral toxicity study in rodents on whole food/feed. EFSA Journal 9(12):2438.

EFSA. 2011b. Guidance on risk assessment of food and feed from genetically modified plants. EFSA Journal 9(5):2150.

EFSA. 2011c. Guidance on the post-market environmental monitoring (PMEM) of genetically modified plants. EFSA Journal 9(8):2316.

EOP (Executive Office of the President). 2016. National Strategy for Modernizing the Regulatory System for Biotechnology Products. Available at https://obamawhitehouse.archives.gov/sites/default/files/microsites/ostp/biotech_national_strategy_final.pdf. Accessed January 31, 2017.

EOP. 2017. Modernizing the Regulatory System for Biotechnology Products: An Update to the Coordinated Framework for the Regulation of Biotechnology. Available at https://obamawhitehouse.archives.gov/sites/default/files/microsites/ostp/2017_coordinated_framework_update.pdf. Accessed January 30, 2017.

EPA (U.S. Environmental Protection Agency). 1998. Guidelines for Ecological Risk Assessment. Federal Register 63:26846–26924.

EPA. 2004. Overview of the Ecological Risk Assessment Process in the Office of Pesticide Programs, U.S. Environmental Protection Agency. Endangered and Threatened Species Effects Determinations. Available at https://www.epa.gov/sites/production/files/2014-11/documents/ecorisk-overview.pdf. Accessed September 14, 2016.

EPA. 2012a. SAP Minutes No. 2012-06: A Set of Scientific Issues Being Considered by the Environmental Protection Agency Regarding Pollinator Risk Assessment Framework. Available at https://www.regulations.gov/document?D=EPA-HQ-OPP-2012-0543-0047. Accessed September 14, 2016.

EPA. 2012b. SAP Minutes No. 2012-05: A Set of Scientific Issues Being Considered by the Environmental Protection Agency Regarding: Problem Formulation for the Reassessment of Ecological Risks from the Use of Atrazine. Available at https://www.epa.gov/sites/production/files/2015-06/documents/061212minutes.pdf. Accessed September 14, 2016.

Exponent. 2005. DEEM-FCID TM Distributional Acute Dietary Exposure Analysis Program, Version 2.03. Exponent, Washington, DC.

FDA (U.S. Food and Drug Administration). 2012. A Guide to Drug Safety Terms at FDA. FDA Consumer Health Information. November. Available at http://www.fda.gov/downloads/ForConsumers/ConsumerUpdates/UCM107976.pdf. Accessed January 9, 2017.

FDA. 2017. Regulation of Mosquito-Related Products; Draft Guidance for Industry; Availability. Federal Register 82:6574–6575.

Forbes, V.E., P. Calow, and R.M. Sibly. 2001. Are current species extrapolation models a good basis for ecological risk assessment? Environmental Toxicology and Chemistry 20(2):442–447.

Freudenburg, W.R. 1996. Strange chemistry: Environmental risk conflicts in a world of science, values, and blind spots. Pp. 11–36 in Handbook for Environmental Risk Decision Making: Values, Perceptions, and Ethics, C.R. Cothern, ed. Boca Raton, FL: Lewis Publishers.

Galdzicki, M., K.P. Clancy, E. Oberortner, M. Pocock, J.Y. Quinn, C.A. Rodriguez, N. Roehner, M.L. Wilson, L. Adam, J.C. Anderson, B.A. Bartley, J. Beal, D. Chandran, J. Chen, D. Densmore, D. Endy, R. Grünberg, J. Hallinan, N.J. Hillson, J.D. Johnson, A. Kuchinsky, M. Lux, G. Misirli, J. Peccoud, H.A. Plahar, E. Sirin, G-B. Stan, A. Villalobos, A. Wipat, J.H. Gennari, C.J. Myers, and H.M. Sauro. 2014. The Synthetic Biology Open Language (SBOL) provides a community standard for communicating designs in synthetic biology. Nature Biotechnology 32(6):545–550.

Gibson, D.G., J.I. Glass, C. Lartigue, V.N. Noskov, R.-Y. Chuang, M.A. Algire, G.A. Benders, M.G. Montague, L. Ma, M.M. Moodie, C. Merryman, S. Vashee, R. Krishnakumar, N. Assad-Garcia, C. Andrews-Pfannkoch, E.A. Denisova, L. Young, Z.-Q. Qi, T.H. Segall-Shapiro, C.H. Calvey, P.P. Parmar, C.A. Hutchison, H.O. Smith, and J.C. Venter. 2010. Creation of a bacterial cell controlled by a chemically synthesized genome. Science 329(5987):52–56.

Ginsberg, W. 2015. Creating a Federal Advisory Committee in the Executive Branch. Washington, DC: Congressional Research Service.

Goodman, R.E., M. Ebisawa, F. Ferreira, H.A. Sampson, R. van Ree, S. Vieths, J.L. Baumert, B. Bohle, S. Lalithambika, J. Wise, and S.L. Taylor. 2016. AllergenOnline: A peer-reviewed, curated allergen database to assess novel food proteins for potential cross-reactivity. Molecular Nutrition and Food Research 60:1183–1198.

Guston, D.H. 2014. Understanding anticipatory governance. Social Studies of Science 44(2):218–242.

Harding, S. 1996. Rethinking standpoint epistemology: What is "strong objectivity"? Pp. 235–248 in Feminism and Science, E.F. Keller and H. Longino, eds. Oxford: Oxford University Press.

Hayes, K.R., B. Leung, R. Thresher, J.M. Dambacher, and G.R. Hosack. 2014. Meeting the challenge of quantitative risk assessment for genetic control techniques: A framework and some methods applied to the common Carp (*Cyprinus carpio*) in Australia. Biological Invasions 16(6):1273–1288.

Iftikhar, S., S. Khan, Z. Anwar, and M. Kamran. 2015. GenInfoGuard—a robust and distortion-free watermarking technique for genetic data. PLOS ONE 10(2):e0117717.

IRGC (International Risk Governance Council). 2015. Guidelines for Emerging Risk Governance. Lausanne, Switzerland: IRGC. Available at https://www.irgc.org/wp-content/uploads/2015/03/IRGC-Emerging-Risk-WEB-31Mar.pdf. Accessed October 17, 2016.

Kapuscinski, A.R., R.M. Goodman, S.D. Hann, L.R. Jacobs, E.E. Pullins, C.S. Johnson, J.D. Kinsey, R.L. Krall, A.G. La Viña, M.G. Mellon, and V.W. Ruttan. 2003. Making "safety first" a reality for biotechnology products. Nature Biotechnology 21:599–601.

Kapuscinski, A.R., S. Li, K.R. Hayes, and G. Dana, eds. 2007. Environmental Risk Assessment of Genetically Modified Organisms, Vol. 3. Methodologies for Transgenic Fish. Cambridge, MA: CABI. Available at https://gmoera.umn.edu/sites/gmoera.umn.edu/files/environmental_risk_assessment_volume_3.pdf. Accessed October 28, 2016.

Krafsur, E.S. 1998. Sterile insect technique for suppressing and eradicating insect populations: 55 years and counting. Journal of Agricultural Entomology 15(4):303–317.

Kuzma, J. 2006. Nanotechnology oversight and regulation—just do it. Environmental Law Reporter 36:10913–10923.

Kuzma, J. 2013. Properly paced or problematic? Examing governance of GMOs. Pp. 176–197 in Innovative Governance Models for Emerging Technologies, G. Marchant, K. Abbott, and B. Allenby, eds. Cheltenham, UK: Edward Elgar.

Kuzma, J., and J.C. Besley. 2008. Ethics of risk analysis and regulatory review: From bio- to nanotechnology. Nanoethics 2(2):149–162.

Kuzma, J., J. Romanchek, and A. Kokotovich. 2008. Upstream oversight assessment for agrifood nanotechnology: A case studies approach. Risk Analysis 28(4):1081–1098.

Lavertu, S., and D.L. Weimer. 2010. Federal advisory committees, policy expertise, and the approval of drugs and medical devices at the FDA. Journal of Public Administration Research and Theory 21:211–237.

Linkov, I., F.K. Satterstrom, J. Steevens, E. Ferguson, and R.C. Pleus. 2007. Multi-criteria decision analysis and environmental risk assessment for nanomaterials. Journal of Nanoparticle Research 9(4):543–554.

Liss, M., D. Daubert, K. Brunner, K. Kliche, U. Hammes, A. Leiherer, and R. Wagner. 2012. Embedding permanent watermarks in synthetic genes. PLOS ONE 7(8):e42465.

Marchant, G.E., and W. Wallach. 2015. Coordinating technology governance. Issues in Science and Technology 31(4):43–50.

McHughen, A., and S. Smyth. 2008. U.S. regulatory system for genetically modified [genetically modified organism (GMO), rDNA or transgenic] crop cultivars. Plant Biotechnology Journal 6(1):2–12.

Meghani, Z., and J. Kuzma. 2011. The "revolving door" between regulatory agencies and industry: A problem that requires reconceptualizing objectivity. Journal of Agricultural and Environmental Ethics 24(6):575–599.

Moffitt, S.L. 2014. Making Policy Public: Participatory Bureaucracy in American Democracy. New York: Cambridge University Press.

Murphy, B., C. Jansen, J. Murray, and P. De Barro. 2010. Risk Analysis on the Australian Release of *Aedes aegypti* (L.) (Diptera: Culicidae) Containing *Wolbachia*. Indooroopilly, Australia: CSIRO.

Murray, J.V., C.C. Jansen, and P. De Barro. 2016. Risk associated with the release of *Wolbachia*-infected *Aedes aegypti* mosquitoes into the environment in an effort to control dengue. Frontiers in Public Health 4:43.

Naranjo, S.E., G. Head, and G.P. Dively. 2005. Field studies assessing arthropod nontarget effects in *Bt* transgenic crops: Introduction. Environmental Entomology 34(5):1178–1180.

NASEM (National Academies of Sciences, Engineering, and Medicine). 2016a. Gene Drives on the Horizon: Advancing Science, Navigating Uncertainty, and Aligning Research with Public Values. Washington, DC: The National Academies Press.

NASEM. 2016b. Genetically Engineered Crops: Experiences and Prospects. Washington, DC: The National Academies Press.

Nelson, K.C., G. Kibata, L. Muhammad, J. Okuro, F. Muyekho, M. Odindo, A. Ely, and J. Waquil. 2004. Problem formulation and options assessment (PFOA) for genetically modified organisms: The Kenya case study. Pp. 57–82 in Environmental Risk Assessment of Genetically Modified Organisms, Vol. 1. A Case Study of Bt Maize in Kenya, A. Hilbeck and D.A. Andow, eds. Cambridge, MA: CABI.

Nielson, A.A., B.S. Der, J. Shin, P. Vaidyanathan, V. Paralanov, E.A. Strychalski, D. Ross, D. Densmore, and C.A. Voight. 2016. Genetic circuit design automation. Science 352(6281):aac7341.

NRC (National Research Council). 1996. Understanding Risk: Informing Decisions in a Democratic Society. Washington, DC: National Academy Press.

NRC. 2007. Models in Environmental Regulatory Decision Making. Washington, DC: The National Academies Press.

NRC. 2008. Public Participation in Environmental Assessment and Decision Making. Washington, DC: The National Academies Press.

NRC. 2009. Science and Decisions: Advancing Risk Assessment. Washington, DC: The National Academies Press.

NRC. 2013. Assessing Risks to Endangered and Threatened Species from Pesticides. Washington, DC: The National Academies Press.

OECD (Organisation for Economic Co-operation and Development). 2014. Guidance Document for Describing Non-Guideline In Vitro Test Methods. Series on Testing and Assessment No. 211. Paris: OECD.

Pauwels, K., S.C. De Keersmaecker, A. De Schrijver, P. du Jardin, N.H. Roosens, and P. Herman. 2015. Next-generation sequencing as a tool for the molecular characterisation and risk assessment of genetically modified plants: Added value or not? Trends in Food Science & Technology 45(2):319–326.

Pennington, K.M., A.R. Kapuscinski, M.S. Morton, A.M. Cooper, and L.M. Miller. 2010. Full life-cycle assessment of gene flow consistent with fitness differences in transgenic and wild-type Japanese medaka fish (*Oryzias latipes*). Environmental Biosafety Research 9(1):41–57.

Peterson, R.K., S.J. Meyer, A.T. Wolf, J.D. Wolt, and P.M. Davis. 2006. Genetically engineered plants, endangered species, and risk: A temporal and spatial exposure assessment for Karner blue butterfly larvae and Bt maize pollen. Risk Analysis 26(3):845–858.

Ramachandran, G., S. Wolf, J. Paradise, J. Kuzma, R. Hall, E. Kokkoli, and L. Fatehi. 2011. Recommendations for oversight of nanobiotechnology: Dynamic oversight for complex and convergent technology. Journal of Nanoparticle Research 13(4):1345–1371.

Raybould, A., G. Caron-Lormier, and D.A. Bohan. 2011. Derivation and interpretation of hazard quotients to assess ecological risks from the cultivation of insect-resistant transgenic crops. Journal of Agricultural and Food Chemistry 59(11):5877–5885.

Renn, O. 2005. Risk Governance: Towards and Integrative Approach. Geneva: IRGC.

Roberts, J.P., S. Stauffer, C. Cummings, and J. Kuzma. 2015. Synthetic Biology Governance: Delphi Study Workshop Report. GES Center Report No. 2015.2. Available at https://research.ncsu.edu/ges/files/2014/04/Sloan-Workshop-Report-final-ss-081315-1.pdf. Accessed October 17, 2016.

Rose, R.I., ed. 2007. White Paper on Tier-Based Testing for the Effects of Proteinaceous Insecticidal Plant-Incorporated Protectants on Non-Target Arthropods for Regulatory Risk Assessments. Environmental Protection Agency and U.S. Department of Agriculture–Animal and Plant Health Inspection Service. Available at https://www.epa.gov/sites/production/files/2015-09/documents/tier-based-testing.pdf. Accessed January 15, 2017.

Ryals, J. 2016. Metabolomics as a High Resolution Phenotyping Tool. Webinar presentation to the National Academies of Sciences, Engineering, and Medicine Committee on Future Biotechnology Products and Opportunities to Enhance Capabilities of the Biotechnology Regulatory System, July 29.

Schein, C.H., O. Ivanciuc, and W. Braun. 2007. Bioinformatics approaches to classifying allergens and predicting cross-reactivity. Immunology and Allergy Clinics of North America 27:1–27.

Sears, M.K., R.L. Hellmich, D.E. Stanley-Horn, K.S. Oberhauser, J.M. Pleasants, H.R. Mattila, B.D. Siegfried, and G.P. Dively. 2001. Impact of Bt corn pollen on monarch butterfly populations: A risk assessment. Proceedings of the National Academy of Sciences of the United States of America 98(21):11937–11942.

Shapiro, B. 2015. How to Clone a Mammoth: The Science of De-Extinction. Princeton, NJ: Princeton University Press.

Stilgoe, J., R. Owen, and P. Macnaghten. 2013. Developing a framework for responsible innovation. Research Policy 42(9):1568–1580.

Stirling, A. 2007. Risk, precaution and science: Towards a more constructive policy debate. EMBO Reports 8(4):309–315.

Storer, N.P. 2003. A spatially explicit model simulating western corn rootworm (Coleoptera: Chrysomelidae) adaptation to insect-resistant maize. Journal of Economic Entomology 96(5):1530–1547.

Suter, G.W., II. 2007. Ecological Risk Assessment, 2nd Ed. Boca Raton, FL: CRC Press.

Thompson, P.B. 2007. Food Biotechnology in Ethical Perspective, 2nd Ed. The International Library of Environmental, Agricultural and Food Ethics Vol. 10. Dordrecht, Netherlands: Springer.

Tsang, M.P., M.E. Bates, M. Madison, and I. Linkov. 2014. Benefits and risks of emerging technologies: Integrating life cycle assessment and decision analysis to assess lumber treatment alternatives. Environmental Science & Technology 48(19):11543–11550.

Warren-Hicks, W.J., and A. Hart. 2010. Application of Uncertainty Analysis to Ecological Risks of Pesticides. Boca Raton, FL: CRC Press.

Weimer, D.L. 2006. The puzzle of private rulemaking: Expertise, flexibility, and blame avoidance in U.S. regulation. Public Administration Review 66(4):569–582.

Weimer, D.L., and L. Wilk. 2016. Allocation of indivisible life-saving goods with both intrinsic and relational quality: The new deceased-donor kidney allocation system. Administration and Society 1–30.

Wilsdon, J., and R. Willis. 2004. See-Through Science: Why Public Engagement Needs to Move Upstream. London: Demos.

Wolt, J.D. 2011. A mixture toxicity approach for environmental risk assessment of multiple insect resistance genes. Environmental Toxicology and Chemistry 30(3):763–772.

Wolt, J.D., and R.K.D. Peterson. 2010. Prospective formulation of environmental risk assessments: Probabilistic screening for Cry1A(b) maize risk to aquatic insects. Ecotoxicology and Environmental Safety 73:1182–1188.

Wolt, J.D., K. Wang, D. Sashital, and C.J. Lawrence-Dill. 2016. Achieving plant CRISPR targeting that limits off-target effects. The Plant Genome 9(3).

Young, G.J., S. Zhang, H.P. Mirsky, R.F. Cressman, B. Cong, G.S. Ladics, and C.X. Zhong. 2012. Assessment of possible allergenicity of hypothetical ORFs in common food crops using current bioinformatics guidelines and its implications for the safety assessment of GM crops. Food and Chemical Toxicology 50:3741–3751.

6

Conclusions and Recommendations

On the basis of its assessment of the trends in biotechnology as of 2016, the likely products of biotechnology in the next 5–10 years, and the current authorities and capabilities of the regulatory agencies, the committee identified a set of broad themes regarding future opportunities for enhancement of the U.S. biotechnology regulatory system:

- The bioeconomy is growing rapidly and the U.S. regulatory system needs to provide a balanced approach for consideration of the many competing interests in the face of this expansion.
- The profusion of biotechnology products over the next 5–10 years has the potential to overwhelm the U.S. regulatory system, which may be exacerbated by a disconnect between research in regulatory science and expected uses of future biotechnology products.
- Regulators will face difficult challenges as they grapple with a broad array of new types of biotechnology products—for example, cosmetics, toys, pets, and office supplies—that go beyond contained industrial uses and traditional environmental release (for example, *Bt* or herbicide-resistant crops).
- The safe use of new biotechnology products requires rigorous, predictable, and transparent risk-analysis processes whose comprehensiveness, depth, and throughput mirror the scope, scale, complexity, and tempo of future biotechnology applications.
- In addition to the conclusions and recommendations from this report, the U.S. Environmental Protection Agency (EPA), the U.S. Food and Drug Administration (FDA), the U.S. Department of Agriculture (USDA), and other agencies involved in regulation of future biotechnology products would benefit from adopting recommendations made by previous National Academies committees related to future products of biotechnology, which are consistent with the findings and recommendations in this report.

In this final chapter, the committee has summarized its major conclusions, organized against the statement of task, and made selected recommendations for next steps to be taken to enhance the capabilities of the U.S. regulatory system to protect human health and the environment.

MAJOR ADVANCES AND NEW TYPES OF PRODUCTS

The committee's statement of task requested that the committee "describe the major advances and the potential new types of biotechnology products likely to emerge over the next 5–10 years." In reviewing the technologies that are currently being explored in industry, academia, and government, the committee concluded:

> **Conclusion 6-1: The scale, scope, complexity, and tempo of biotechnology products are likely to increase in the next 5–10 years. Many products will be similar to existing biotechnology products, but they may be created through new processes, and some products may be wholly unlike products that exist today.**

Driven by advances in biotechnology such as lowered cost of gene synthesis and sequencing, intelligent design tools for building gene constructs, readily accessible and affordable standardized biological parts, and the increasing accessibility of those technologies to a broad array of agents, the committee anticipates a significant profusion of biotechnology products in the next 5–10 years. Many of these products will be for contained use or otherwise map to existing product categories, but the committee expects that there will be significantly more products designed to exist in the open environment under low-management conditions that will be also developed. Some of these products, like the mosquito engineered to produce sterile offspring (deployment of which was under discussion in Florida at the time the committee was writing its report), will be designed to mitigate an identified environmental problem, a public health issue (for example, reducing populations of mosquitoes that transmit malaria, dengue, and Zika), or both. In addition, new processes for making genetic modifications will be introduced, moving from *Agrobacterium*-mediated transformation to biolistics to genome-editing techniques such as CRISPR, and beyond this to novel advances in genome engineering, including the possibility of genomically recoded organisms that are largely or wholly created via synthetic DNA.

These advances in technology will lead to new products that range from an expansion of the familiar set of organismal hosts and genetic pathways to those that use rapid design-build-test-learn cycles to enable more complex designs of genetic pathways in a wider variety of host organisms to those in which multiple organisms may be used in complex microbial communities, such as microbiome engineering. As biotechnology products become less similar to existing products and more complex, there will be fewer comparator products to use in risk analyses; consequently, there will be a need to develop new approaches to analyze risks. The number of actors who are involved in product development will also increase and diversify beyond just companies and academia, creating the additional regulatory challenge of product developers who may have little familiarity with risk analysis and with the regulatory system.

CURRENT RISK-ANALYSIS SYSTEM AND AGENCY AUTHORITIES

The committee was also asked to "describe the existing risk analysis system for biotechnology products including, but perhaps not limited to, risk analyses developed and used by EPA, USDA, and FDA, and describe each agency's authorities as they pertain to the products of biotechnology." In carrying out this charge, the committee reviewed the Coordinated Framework for Regulation of Biotechnology, including the 2017 update to the Coordinated Framework (EOP, 2017) as well as the 2016 *National Strategy for Modernizing the Regulatory System for Biotechnology Products* (EOP, 2016). The committee reached the following conclusions:

Conclusion 6-2: The Coordinated Framework for Regulation of Biotechnology appears to have considerable flexibility to cover a wide range of biotechnology products, though in some cases the jurisdiction of the agencies has the potential to leave gaps in regulatory oversight.

The Coordinated Framework for Regulation of Biotechnology is a complex collection of statutes and regulations that provides the basis for federal oversight of biotechnology products. The Coordinated Framework appears to have considerable flexibility to cover a wide range of biotechnology products, although in some cases the agencies' jurisdiction has been defined in ways that potentially may leave gaps in regulatory oversight. Even when the statutes technically do allow agencies to regulate these products, the current statutes equip the regulators with tools that may, at times, make it hard for them to regulate the products effectively. For example, the statutes may not empower regulators to require product developers to share in the burden of generating information about product safety, may place the burden of proof on regulators to demonstrate that a product is unsafe before they can take action to protect the public, or may require cumbersome processes or procedures the regulators must follow before they can act. Almost all of the statutes lack adequate legal authority for post-marketing surveillance, monitoring, and continuous-learning approaches. Thus, although the products of future biotechnology often are within the jurisdiction of existing regulators, they will struggle to regulate these products effectively and to respond nimbly to the profusion of products that will be coming.

Conclusion 6-3: The current biotechnology regulatory system is complex and fragmented, resulting in a system that can be difficult for individuals, nontraditional organizations, and small- and medium-sized enterprises to navigate, that might cause uncertainty and a lack of predictability for developers of future biotechnology products, and that has the potential for loss of public confidence in regulation of future biotechnology products.

As was pointed out in the public comments received as part of developing the 2017 update to the Coordinated Framework, the U.S. regulatory system can be difficult to navigate. Future product developers will include new players, such as do-it-yourself biotechnology enthusiasts, nontraditional manufacturers, and entities whose research or product development is funded through non-traditional sources. Protecting public safety may, at times, call for controls over who can access and use certain types of products—for example, to restrict the use of the product to qualified users or to ensure the product is used only in facilities that agree to implement certain safety measures. The regulatory agencies have little authority to restrict sales, distribution, and use of products that do not cross state lines or are made and used in domestic settings (individuals' homes). The federal frameworks currently in place for limiting access to biological agents—the Federal Select Agent Program administered by USDA's Animal and Plant Health Inspection Service (APHIS) and the Centers for Disease Control and Prevention (CDC), the U.S. Department of Commerce's restrictions on transactions, and voluntary screening programs administered by the U.S. Department of Health and Human Services—are presently geared to controlling small numbers of highly dangerous or strategically significant products, rather than a wider array of biotechnology products that may require qualified users in order to be safe. In addition, many developers of early-stage biotechnology products or biological technology (that may eventually lead to products for commercial use) do not consider regulatory issues and the potential need for data to support risk analyses during technology (and sometimes product) development. Failure to anticipate regulatory requirements creates the potential to complicate developers' business plans and delay the risk analyses associated with the decisions to allow future biotechnology products to enter the market.

FUTURE PRODUCTS UNDER THE CURRENT COORDINATED FRAMEWORK

Regarding future products of biotechnology, the committee was asked to "determine whether potential future products could pose different types of risks relative to existing products and organisms. Where appropriate, identify areas in which the risks or lack of risks relating to the products of biotechnology are well understood." As described in Chapter 4, in carrying out this portion of its statement of task, the committee distinguished between risk-assessment endpoints and the complexity of risk assessments and also interpreted "well understood" to mean that the uncertainty in estimates of risk does not preclude a formulation of risk-management options. The committee reached the following conclusions:

Conclusion 6-4: The risk-assessment endpoints for future biotechnology products are not new compared with those that have been identified for existing biotechnology products, but the pathways to those endpoints have the potential to be very different in terms of complexity.

The biotechnology products emerging in the next 5–10 years pose a diverse array of environmental, health, and safety risks that vary widely in terms of their potential impacts, likelihood of occurrence, spatial and temporal dimensions, and the appropriate regulatory policies for their assessment. Although the nature of human health and environmental risk-assessment endpoints that will need consideration are similar to those identified with existing products, the pathways to these endpoints will differ in complexity; therefore, advances in regulatory science will be needed for effective and appropriate evaluation. The number of products poised to enter the marketplace in the coming years will likely outpace the means and capacity for voluntary- or regulatory-based assessment processes to inform decision making. This imbalance, if not addressed in the near term, could impede the development of new biotechnology products in the long term. The profusion of future biotechnology products poses a significant potential stress to the existing regulatory system. Regulatory agencies are likely not prepared with sufficient staff, appropriate risk-assessment approaches, and corresponding guidance for development and evaluation of product data packages.

Conclusion 6-5: The profusion of future biotechnology products anticipated in coming years will challenge the federal agencies' ability to handle significant increases in the rate of biotechnology product innovation, the number of biotechnology products, the complexity of interactions, and the diversity of actors (and their experience with the regulatory process).

Based on the information gathered as part of this study, there is a strong possibility that the number of products per year that will require federal oversight will increase and the complexity of future assessments for these products, and the associated level of effort required on the part of appropriate regulatory authorities, will also increase. The committee saw this not as a transient event at this point in time, but rather as part of a sustained increase in the development of new products of biotechnology that will be driven by increased understanding of the biological sciences and increased capabilities in the underlying biological engineering technologies. New tools for government oversight in the face of this expansion may be required.

Conclusion 6-6: To enable effective regulation of the safe use of future biotechnology products, it would be beneficial to have a single point of entry into the regulatory system with a decision-making structure aimed to assess and manage product risk, to direct

products to their appropriate regulatory agencies, and to increase transparency for developers and society.

In order to be prepared for the potential profusion of future products of biotechnology, it will be important that the U.S. regulatory system has the capacity and capability to efficiently assess their potential human health and environmental risks. A revised strategy for regulatory oversight with increased public participation, transparency, and predictability, when possible under current statutes, can improve public confidence in the regulatory process for future biotechnology products. A rigorous, transparent regulatory oversight system can ensure that risk-analysis efforts are proportional to the level of regulatory agencies' familiarity with the product and the degree of complexity required in the associated risk assessment. The need to balance the many different demands from federal agencies, developers, and interested and affected parties will make the current case-by-case assessment of new products of biotechnology increasingly challenging. Furthermore, it will be important that the routes of decision making be consistently applied across the different classes of products described in Chapter 2. The committee notes that the ability to provide a consistent regulatory route that depends on product function and use may be limited by the jurisdiction of the agencies and the differences in the authorities that govern them.

As described in the 2017 update to the Coordinated Framework and the *National Strategy*, individual federal agencies have already taken significant steps toward providing information to a variety of stakeholders in their individual processes. The committee concluded that, in addition to these initial steps, a single point of entry that allows better understanding and selection of the regulatory route would further enhance the ability of the regulatory system to handle anticipated products of biotechnology and the diverse array of developers of such products.

OPPORTUNITIES FOR ENHANCEMENT

Finally, the committee was asked to "indicate what scientific capabilities, tools, and expertise may be useful to the regulatory agencies to support oversight of potential future products of biotechnology." The committee reviewed the current expertise within EPA, FDA, and USDA and analyzed the investments that those agencies and others are making in areas related to risk analyses. The committee concluded:

Conclusion 6-7: The staffing levels, expertise, and resources available in EPA, FDA, USDA, and other agencies that have interests related to future biotechnology products may not be sufficient to handle the expected scope and scale of future biotechnology products.

Although the regulatory agencies have access to a number of external advisory committees, the number of in-house experts and the responses to the committee's request for information indicate that there may not be sufficient scientific capability, capacity, and tools within and across the agencies to address the risk-assessment challenges for future biotechnology products. The agencies need to maintain or build the capacity and the level of expertise within agencies required to assess the anticipated profusion of biotechnology products and have access to the most current tools for technology assessment. Some resources may also need to go into research. The specific areas of scientific capabilities, tools, and expertise that are likely to be required are described in the committee's recommendations in the next section.

RECOMMENDATIONS

On the basis of the conclusions of this report, the committee recommends the following actions be taken to enhance the ability of the biotechnology regulatory system to oversee the consumer safety and environmental protection required for future biotechnology products.

Recommendation 1: EPA, FDA, USDA, and other agencies involved in regulation of future biotechnology products should increase scientific capabilities, tools, expertise, and horizon scanning in key areas of expected growth of biotechnology, including natural, regulatory, and social sciences.

Future biotechnology products will be more complex in terms of their internal and external interactions, and it is critical that the agencies involved in regulation of biotechnology develop and maintain scientific capabilities, tools, and expertise in relevant areas. Furthermore, it will be essential that the agencies stay apprised of technology trends so that they can engage in meaningful discussions with technology and product developers early in the product-development cycle, where there is often the best opportunity to affect future technologies. Thus, it will be also important to maintain the capability to perform horizon scanning through participation in technical meetings, outreach to universities and companies, and engagement with the public and international partners.

Determination of the key areas of scientific capability will need to adapt to the emerging technologies that underlie future products of biotechnology. On the basis of the current level of federal investments, some of the key research areas in regulatory science related to the products of biotechnology likely in the next 5–10 years include comparators, off-target gene effects, and phenotypic characterization; gene fitness, genetic stability, and horizontal gene transfer; impacts on nontarget organisms; control of organismal traits; modeling (including risk-analysis approaches under uncertainty) and life-cycle analyses; monitoring and surveillance; and economic and social costs and benefits.

In maintaining an active list of relevant areas of expertise that should be developed and maintained, federal agencies should partner with or otherwise engage the developer sector to ensure that the regulatory-science priorities are informed by the nature of biotechnology products in discovery, research, and development pipelines.

To support this broad recommendation, the committee developed several more specific recommendations on opportunities for increasing the capabilities of the regulatory system.

Recommendation 1-1: Regulatory agencies should build and maintain the capacity to rapidly triage products entering the regulatory system that resemble existing products with a history of characterization and use, thus reducing the time and effort required for regulatory decision making, and should be prepared to focus questions on identifying new pathways to risk-assessment endpoints associated with products that are unfamiliar and that require more complex risk assessments.

To encourage innovation, execute diligence in risk analyses, and instill confidence in the process and outcomes of regulating future products of biotechnology, it is important that regulatory agencies make use of a stratified approach to identify those products that require the most attention. Although the risk-assessment endpoints from biotechnology products have been considered in the past and existing processes for risk assessment can be applied, processes need to be advanced to provide quantitative risk estimates and address the dimensionality of more complex assessments. In some areas, decades of work and review of multiple products are available, and future products that are similar to those with a history of characterization and use should be able to take advantage of these existing risk analyses to bridge existing information and focus exposure or hazard data to specific, identified areas of uncertainty unique to a new product. It will be important to focus resources on those products of biotechnology in which novel traits and more complex interactions are present because these are the types of products for which detailed scientific understanding may not yet be available and with which regulatory agencies may be less familiar. Even in these situations, risk-analysis frameworks should encourage the use of clearly articulated conceptual models based on best available information to identify critical exposure pathways and risk-assessment endpoints to guide information needs. In some cases, an iterative approach to information gathering can support a more efficient and focused risk analysis (for example, the approach described in the National Academies report on gene drives [NASEM, 2016a]).

It is likely that increased staffing, training, and expertise in new disciplines will be needed to deal with the coming profusion of new biotechnology products to support timely regulatory decisions that are based on the best available science tailored to the complexity of a given risk analysis. Analysis of capabilities in the regulatory agencies over the period 2011–2015 indicates that in some cases there is a decrease in the number of personnel available for nonhealth-related regulatory activities.

Recommendation 1-2: In order to inform the regulatory process, federal agencies should build capacity to scan the horizon for new products and processes that could present novel risk pathways, develop new approaches to assess and address more complex risk pathways, and implement mechanisms for keeping regulators aware of the emerging technologies they have to deal with.

In order to be prepared for future products of biotechnology that will enter the regulatory pipeline, regulatory agencies should have an informed view of the underlying technologies that will lead to those advancements. Regulatory agencies should provide training programs to continually maintain technical expertise commensurate with the scope, scale, and complexity of future biotechnology products and may wish to consider annual or biannual training at technical workshops to learn and apply state-of-the-art technologies. Regulatory agencies should also critically assess existing expertise and develop a strategic plan to hire personnel (permanent or contract employees) capable of assessing future biotechnology products.

In addition to building and maintaining internal expertise, regulatory agencies should make use of external resources for horizon scanning. This could include the use of external advisory groups such as EPA's Science Advisory Board and Scientific Advisory Panel, and extramural research to identify and study emerging risks. EPA's futures network may also provide insights into new methods for effective horizon scanning. Agencies such as the U.S. Department of Energy, the U.S. Department of Defense, the National Institute of Standards and Technology, the National Aeronautics and Space Administration, and the National Science Foundation could assist the regulatory agencies in their horizon-scanning efforts.

Recommendation 1-3: EPA, FDA, USDA, and other relevant federal agencies should work together to (1) pilot new approaches for problem formulation and uncertainty characterization in ecological risk assessments, with peer review and public participation, on open-release products expected during the next 5 years; (2) formulate risk–benefit assessment approaches for future products, with particular emphasis on future biotechnology products with unfamiliar functions and open-release biotechnology products; and (3) pool skills and expertise across the government as needed on first-of-a-kind risk–benefit cases.

There is a significant amount of expertise available across the various federal agencies involved with the regulation of biotechnology products, and the increasingly complex interactions between traits, functions, and the environment that are likely with future products of biotechnology motivate increased collaboration between the agencies. A potential mechanism to increase cooperation would be to use a "community of practice"—comprised of representatives from the agencies' community of risk assessors and risk managers—that could explore approaches to advance science-based assessments of new products, especially in those areas where products cut across traditional areas of regulatory oversight by individual agencies. As an example, the U.S. Fish and Wildlife Service has a conservation genetics community of practice with a mission "to serve as an interactive forum to facilitate the growth, application and exchange of conservation genetics expertise, information, and technology among members" (FWS, 2011). Regulators of biotechnology products could form a similar community of practice to share and build knowledge about new biotechnology processes and new ways to evaluate the risks of biotechnology products.

In developing this "community of practice," the federal agencies could consider the entire life cycle of biotechnology products as well as approaches to engage interested and affected parties in governance and oversight of risk analysis and regulation of those products. Risk analyses will vary by statutory requirements, the regulatory agencies' familiarity with the product, and manufacturing processes and use patterns of the products; hence, governance and oversight approaches should be tailored to the regulatory context. Federal agencies should explore the implementation of principles of responsible research and innovation and should provide public confidence in development and use of new biotechnology products.

In many cases, biotechnology-sector regulators need to consider not only the risks posed by future products but also the potential societal benefits that may be gained with a new product. Working through a "community of practice," regulatory agencies could explore opportunities to apply their discretionary powers to focus information needs tailored to products and use patterns, which will inform timely and robust risk–benefit analyses and regulatory decisions. For example, agencies may choose an area of current technology investment (for example, engineered microbial consortia) and begin to examine scenarios and seek external input that would inform future risk analyses.

The agencies could also consider expanding a "community of practice" to include product developers, risk-analysis consultants, state agencies, and other interested and affected parties. This approach has been employed by EPA for other areas of environmental risk assessment and research, as noted in Chapter 4.[1]

> **Recommendation 1-4: EPA, FDA, USDA, and other relevant federal agencies should create a precompetitive or preregulatory review "data commons" that provides data, scientific evidence, and scientific and market experience for product developers.**

A key element in future regulation of biotechnology products will be increased scientific understanding of the complex interactions between a variety of traits and functions that will form the core of future products of biotechnology and the increasingly complex interactions between products of biotechnology and their host environments. To accelerate the advancement of this scientific understanding, appropriate "data commons" could be established and run by the federal government, perhaps based at a national laboratory. These data commons could make use of data produced by government, industry, or academic researchers at a variety of stages of product devel-

---

[1]See, for example, Computational Toxicology Communities of Practice, available at https://www.epa.gov/chemical-research/computational-toxicology-communities-practice, and Pesticide Environmental Modeling Public Meeting–Information, available at https://www.epa.gov/pesticide-science-and-assessing-pesticide-risks/environmental-modeling-public-meeting-information. Both accessed January 10, 2017.

opment. Examples of existing databases include Allergen Online, the International Life Sciences Institute crop composition database, and the CRISPR Genome Analysis Tool.

If properly developed, these data commons could have multiple uses that benefit the capability and capacity of the regulatory system. Regulators could use the data commons as a source of comparative data and quantitative measures of long-term effects. Applicants, especially small- and medium-sized enterprises and entrepreneurs, could use the shared data commons to minimize burdens, promote quality control, and speed up their "time to market" and understanding of societal benefits of biotechnology. Researchers could both tap into these resources to identify gaps in understanding and conduct new studies that enrich the shared pool of knowledge.

Funding for these resources could rely on industry funding or modest user fees to support their development and curation.

> **Recommendation 1-5: Consistent with the goals and guidance stated by the Office of Science and Technology Policy in the Executive Office of the President in a July 2015 memo, the Biotechnology Working Group should implement a more permanent, coordinated mechanism to measure progress against and periodically review federal agencies' scientific capabilities, tools, expertise, and horizon scanning as they apply to the profusion of future biotechnology products.**

In July 2015, the Biotechnology Working Group (BWG) under the Emerging Technologies Interagency Policy Coordination Committee was established with representatives from the Executive Office of the President, EPA, FDA, and USDA (EOP, 2015). The BWG developed the 2017 update to the Coordinated Framework to clarify the current roles and responsibilities of the agencies that regulate the products of biotechnology (EOP, 2017); it also wrote the *National Strategy* (EOP, 2016). In the *National Strategy*, the BWG is charged with producing an annual report on specific steps that agencies are taking to implement that strategy and any other steps that the agencies are taking to improve the transparency, coordination, predictability, and efficiency of the regulation of biotechnology products.

The committee supports the initial steps of the BWG and encourages it to establish more permanent mechanisms to measure progress against and periodically review federal agencies' scientific capabilities, tools, expertise, and horizon scanning as they apply to the profusion of future biotechnology products.

> **Recommendation 2: EPA, FDA, USDA, and other relevant agencies should increase their investments in internal and external research and their use of pilot projects to advance understanding and use of ecological risk assessments and benefit analyses for future biotechnology products that are unfamiliar and complex and to prototype new approaches for iterative risk analyses that incorporate external peer review and public participation.**

Risk governance of future biotechnology products must be capable of adapting and responding to the rapid pace of changes in technology and information that are driving the development of those products. Just as the design-build-test-learn cycle is an important trend that is driving much of the biotechnology field, the use of pilot projects as a means to explore new methods of performing scientifically driven risk analysis can provide an opportunity to rapidly explore the "design space" of regulatory science relevant to future products of biotechnology.

Examples of areas where pilot projects should be considered are given in the detailed recommendations below. The recommendations highlight some of the broad areas where new approaches might be particularly fruitful, but they are not intended to be exhaustive. One key area where the committee believes that advances can be fruitful is in new techniques for iterative risk assessments that link to the design-build-test-learn cycle and also to the scaled release of biotechnology products (from laboratory scale to small field trials, larger field trials, and eventually full-scale deployment). This conceptual approach has been articulated for decades, but the employment of field studies and computational models to inform risk assessments at the population, community, or ecosystem level is often not used or, when used, has generally failed to reduce uncertainties in risk assessments and in some cases raised more questions than answers. In this regard, pilot efforts could be undertaken to evaluate, and develop new approaches as appropriate, open-release products, using physical models (for example, mesocosms or field studies) and computational models to assess population, community, and ecosystem effects.

Another is the use of stratified approaches that focus the most attention on those products for which there is the least familiarity within the regulatory agencies and for which the most complexity is required in the associated risk assessment. These approaches are common practice for regulation of open-release products but could be applied more broadly to contained products and biotechnology platforms. Pilot efforts could be used to develop new, high-throughput comparative risk-assessment methods to ensure a risk-based approach that identifies products with the greatest potential of risk but does not inadvertently become a bottleneck in the regulatory pipeline.

In the area of external participation in the regulatory process, pilot projects can also be used to explore new ways of incorporating expert advice through peer review and providing public participation and input in the regulatory decision-making process. Issues to be addressed through the pilots might include how external, independent peer reviewers are chosen, how the public is identified and engaged, who decides the external participants, and how the external, independent peer review and public participation are integrated into the decision-making process. It may also be possible to further explore how early engagement of external input affects the pace of regulatory approval by encouraging more deliberation in the early stages of the regulatory process rather than post-decision litigation. An example of existing pilot programs that is representative of what could be done is the long-running *Pseudomonas fluorescens* bioremediation project (Trögl et al., 2012; Ji et al., 2013), which could be used to drive future pilots for release of biosensors, modified consortia, and other similar products. Another example is the case of genetically engineered algae and the launch of EPA's algal biotechnology project (EPA, 2015).

To support this broad recommendation, the committee developed several more specific recommendations on potential pilot projects that might be considered.

Recommendation 2-1: Regulatory agencies should create pilot projects for more iterative processes for risk assessments that span development cycles for future biotechnology products as they move from laboratory scale to prototype or field scale to full-scale operation.

The pilot projects could address iterative assessments within design cycles for future biotechnology products by adapting the approach outlined in the National Academies report on gene drives (NASEM, 2016a), which described a path from product-development research through open environmental release with post-release monitoring. Risk analyses could inform decisions at each of five steps in an idealized development scheme that includes preparation for research (step 1), laboratory-based research (step 2), pilot-plant manufacturing or field-based research (step 3), staged market entry or environmental release (step 4), and post-market or post-release surveillance, when appropriate (step 5). Although described as a linear process, development cycles typically involve feedback loops with refined understanding based on new findings and data generated during the course of product research and risk assessments. These pilot projects would benefit from active participation of product developers and interested and affected parties to explore approaches for efficiently integrating risk analyses and options for public engagement within product-development cycles.

Recommendation 2-2: Government agencies should pilot advances in ecological risk assessments and benefit analyses for open-release products expected in the next 5–10 years, with external, independent peer review and public participation.

The biotechnology products emerging in the next 5–10 years pose a diverse array of potential environmental risks that vary widely in terms of their potential impacts, likelihood of occurrence, spatial and temporal dimensions, and the appropriate regulatory policies for their assessment. Although the nature of environmental risk-assessment endpoints that will need consideration are similar to those identified with existing biotechnology products, the pathways to these endpoints will differ in complexity. Regulatory agencies are likely not prepared with sufficient staff, appropriate ecological risk-assessment approaches, and corresponding guidance for development and evaluation of associated product data packages. Public confidence in government oversight of emerging technologies may be eroded to the extent there is a lack of transparency and clarity as to how regulatory authorities are undertaking ecological risk assessments, including identifying societal values in addition to taking input from biotechnology developers in formulating regulatory decisions. Possible pilot efforts could address open release of bioengineered microorganisms or microbial consortia with multiple modifications (for example, for use in bioremediation). Other pilots could address risk assessments for biocontrol agents and the release of non-native organisms (bioengineered or otherwise) designed to suppress and or enhance a species, which reflect a high degree of dimensionality and entail a diversity of risk-assessment endpoints at varying levels of biological organization and temporal scales. Considerations of this scale of complexity may necessitate rethinking of both regulatory processes and risk-assessment approaches.

Recommendation 2-3: Government agencies should initiate pilot projects to develop probabilistic estimates of risks for current products as a means to compare the likelihood of adverse effects of future biotechnology products to existing biotechnology and nonbiotechnology alternatives.

There are many opportunities for increased use of quantitative analyses in risk assessment and, in particular, the use of probabilistic estimates of risks as part of larger conceptual models. Even when data are missing, expert and stakeholder elicitation can be used to identify high-priority areas (Murphy et al., 2010; Hayes et al., 2014; Murray et al., 2016). Such analyses would help identify high-priority information needs to reduce uncertainty in risk estimates and inform the classification of comparable products (including possible alternatives that do not rely on biotechnology approaches) based on the nature of risk-assessment endpoints, dimensionality of risk assessments, and the probabilities of adverse effects. The scenarios discussed in Chapter 4 could be used as a starting point for selecting possible pilot projects. Pilots would be particularly helpful for products intended for wide-area environmental release in low-management conditions.

Recommendation 2-4: Regulatory agencies should make use of pilot projects to explore new methods of outreach to the public and developer community as a means of horizon scanning, assessing need areas for capability growth, and improving understanding of the regulatory process.

One example of a potential pilot in this area would be for one or more regulatory agencies to follow the lead of the Federal Bureau of Investigation (FBI) in making outreach part of its activities to help small companies, the international Genetically Engineered Machine (iGEM) teams, the do-it-yourself biology (DIYbio) community, and others to better understand the regulatory process.

Policy, decision, and social-science research to improve risk-analysis processes is also needed. Research areas could include

1. Experiments with governance systems to test ways to anticipate and prepare for future technologies in governance systems with side-by-side comparisons of different features for these systems and to explore alternatives for engaging interested and affected parties within these systems;
2. Methods to handle uncertainty and ambiguity in governance to improve upstream methods for exploring a broad range of harms and benefits, to characterize uncertainty, and to improve decision-science and future-studies approaches (for example, scenario planning, Bayesian approaches, or systems mapping) in governance systems; and
3. Improved methods to explore claims and counterclaims in contested areas to develop balanced and more inclusive approaches for determining "weight of evidence" to understand and mitigate bias in interpretations of evidence, to acknowledge values behind multiple perspectives and interpretations of evidence, and to explore assumptions, contradictions, and correlation arguments on multiple sides of controversies.

Recommendation 2-5: EPA, FDA, and USDA should engage with federal and state consumer- and occupational-safety regulators that may confront new biotechnology products in the next 5–10 years and make use of pilot projects, interagency collaborations, shared data resources, and scientific tools to pilot new approaches for risk assessment that ensure consumer and occupational safety associated with new biotechnology products, particularly those that may involve novel financing mechanisms, means of production, or distribution pathways.

As described in Chapters 3 and 4, biotechnology regulators will face difficult challenges as they grapple with new product categories that go beyond contained industrial uses and traditional environmental release (for example, *Bt* or herbicide-resistant crops). It will be necessary to engage with other federal agencies to develop appropriate regulatory oversight frameworks for such products. The interagency dialogue should primarily engage agencies such as the Consumer Product Safety Commission, the Occupational Safety and Health Administration, and CDC that have broad responsibility for consumer and occupational safety and public health, but the diversity of future biotechnology products may, on a case-by-case basis, implicate the safety oversight roles of a much longer list of agencies (for example, the National Highway Traffic Safety Administration for automotive applications) that also need effective pathways for engagement in the Coordinated Framework.

Rapid post-marketing risk identification, analysis, and safety surveillance are crucial tools to ensure consumer and occupational safety in an environment that promotes rapid innovation. Agencies should aggressively explore options (for example, public–private partnerships, consumer engagement mechanisms, or anonymous reporting mechanisms such as those that have been effective in airline safety) to enhance timely information flows about the safety of biotechnology products in real-world use.

In addition, as nontraditional research funding mechanisms and do-it-yourself or small-scale uses of biotechnology challenge traditional regulatory enforcement mechanisms, the regulatory agencies could maximize their impact on public safety through their powers to publicize risks and educate the public, their power to convene public workshops and to engage stakeholders in addressing emerging problems, and their power to promote compliance with good practices and codes of conduct via policies that link enforcement discretion to voluntary compliance.

Finally, the safety of some future biotechnology products may call for restrictions on sale, distribution, and use that EPA, FDA, and USDA presently lack statutory authority to implement. The agencies should engage with other concerned agencies (the Federal Select Agent Program administered by USDA–APHIS and CDC, the U.S. Department of Commerce's restrictions on transactions, and voluntary screening programs administered by the U.S. Department of Health and Human Services), with law enforcement (for example, FBI), and with Congress to explore appropriate solutions.

Recommendation 3: The National Science Foundation, the U.S. Department of Defense, the U.S. Department of Energy, the National Institute of Standards and Technology, and other agencies that fund biotechnology research that has the potential to lead to new biotechnology products should increase their investments in regulatory science and link research and education activities to regulatory-science activities.

There are substantial opportunities for the use of improved methods for scientific evaluation, risk assessment, and community engagement related to future products of biotechnology. In order to ensure that the regulatory framework is able to make use of the best available tools in performing its regulatory oversight responsibilities, it will be important to invest in those tools and make them available to regulators and product developers. Areas for consideration include stochastic methods, advances in uncertainty analysis, better ways to integrate and interpret both qualitative and quantitative data, and communication strategies.

In talking with small companies, university research, the DIYbio community, and others during its information-gathering activities, a common theme the committee heard was that early developers of technology that may lead to future biotechnology products are not considering the possible regulatory paths that their technologies and potential products may face. To help address this, federal agencies that fund basic and applied research related to biotechnology should consider funding research activities that close gaps and provide linkages to market-path requirements for regulatory success.

Finally, for those agencies that have a role in educational activities, it may be beneficial to find ways to increase the education that researchers receive regarding the regulatory system. In the same way that the National Institutes of Health (NIH) requires all universities that receive NIH funding to offer courses in biosafety and bioethics, the profusion of future biotechnology products would benefit from the inclusion of material on the regulatory process and risk-assessment frameworks and tools in textbooks, scientific literature, and regulatory agency websites.

To support this broad recommendation, the committee developed several more specific recommendations on how such investments might be targeted.

Recommendation 3-1: The federal government should develop and implement a long-term strategy for risk analysis of future biotechnology products, focused on identifying and prioritizing key risks for unfamiliar and more complex biotechnology products, and work to establish appropriate federal funding levels for sustained, multiyear research to develop the necessary advances in regulatory science.

As noted in Chapter 4, of the total $1.04 billion invested in biotechnology research during 2008–2015, federal research funding agencies invested approximately 7 percent on risk research. The federal government should establish a research program that is responsive to the nature and extent of future challenges facing public- and private-sector risk assessors, risk managers, and other interested and affected parties.

The committee's initial perspective on the highest priorities for increased investments is

- Comparators, off-target gene effects, and phenotypic characterization.
- Genetic fitness, genetic stability, and horizontal gene transfer.
- Impacts on nontarget organisms.
- Control of organismal traits.
- Modeling (including risk-analysis approaches under uncertainty) and life-cycle analyses.
- Monitoring and surveillance.
- Economic and social benefits and costs.

The federal government in coordination with developers should, in an open and transparent process, begin discussions on approaches to establish standards for assay methods, data, and information management. The national laboratories or public–private partnerships may be possible avenues to advancing activities in this area: the U.S. Department of Energy's Bioenergy Research Centers and the Joint Initiative for Metrology in Biology supported by the National Institute of Standards and Technology are examples of relevant entities that are supported by similar strategic federal investments.

Finally, the committee notes that the financial resources needed to establish an adequate research portfolio for the United States need not fall solely on the federal government and the nation's tax payers. The federal government should explore establishing open and transparent approaches to integrate and optimize public investments, private investments, and public–private partnerships to realize the needed resources to support development of a responsive risk-analysis paradigm.

The committee notes that this recommendation is supported by recommendations in the National Academies reports on gene drives and genetically engineered crops, which stated:

> There is an urgent need for publicly funded research on novel molecular approaches for testing future products of genetic engineering so that accurate testing methods will be available when the new products are ready for commercialization. (NASEM, 2016b:236)

> Researchers, regulators, and other decision makers should use ecological risk assessment to estimate the probability of immediate and long-term environmental and public health effects of gene-drive modified organisms and to inform decisions about gene drive research, policy, and applications. (NASEM, 2016a:128)

> To strengthen future ecological risk assessment for gene-drive modified organisms, researchers should design experimental field trials to validate or improve cause-effect pathways and further refine ecological models. (NASEM, 2016a:128)

Recommendation 3-2: Federal agencies that fund early-stage biotechnology-related research and regulatory agencies should provide support to academic, industry, and government researchers to close gaps and provide linkages to market-path requirements for regulatory success.

Many future products of biotechnology build on advances from the research community that take place in environments that are far removed from a regulatory context. Nonetheless, early consideration of regulatory needs could be highly beneficial and could lead to collection of data and incorporation of features that provide substantial benefit to regulatory science and regulatory activities for products that build on that technology. Examples might include the development of quantitative methodologies for risk assessment in biological products that involve consortia of engineered organisms, where much of the underlying understanding of long-term behavior is currently missing.

The committee notes that this recommendation is supported by recommendations in the National Academies reports on gene drives and on the industrialization of biology, which stated:

> To facilitate appropriate interpretation of the outcomes of an ecological risk assessment, researchers and risk assessors should collaborate early and often to design studies that will provide the information needed to evaluate risks of gene drives and reduce uncertainty to the extent possible. (NASEM, 2016a:128)

> Government agencies, including EPA, USDA, FDA, and NIST, should establish programs for both the development of fact-based standards and metrology for risk assessment in industrial biotechnology and programs for the use of these fact-based assessments in evaluating and updating the governance regime. (NRC, 2015:109)

Recommendation 3-3: Government agencies that fund biotechnology development, working together with regulatory agencies and each other, should also invest in new methods of understanding ethical, legal, and social implications associated with future biotechnology products.

As noted in Chapter 4, research on the ethical, legal, and social implications (ELSI) of biotechnology represented less than 1 percent of total in investment during 2008–2015. It is likely that the ethical, legal, and social implications of future products of biotechnology will increase, so it will be important for federal agencies to ensure that investments in new methods of understanding ELSI associated with biotechnology are appropriately determined. This is especially important in the context of risk analysis as values are embedded in the choices of models, endpoints, risks assessed, and methods used. Investments should include adequate funding to proactively address ethical, legal, and social implications linked to new product introductions. Possible agencies that could support such efforts include the National Science Foundation and the U.S. Department of Energy, and also EPA, FDA, and USDA if they are appropriated funds for biotechnology research.

Recommendation 3-4: Government agencies with an educational mission should identify and fund activities that increase awareness and knowledge of the regulatory system in courses and educational materials for students whose research will lead to advances in biotechnology products.

At the time of completion of their degrees and postgraduate training, many of the students and postdocs who will go on to make important contributions to the technology of future biotechnology products may not be aware of the Coordinated Framework and may not have been exposed to methods of risk assessment and risk management that are part of the Coordinated Framework. Encouraging the development and inclusion of educational materials that provide insights and context to the regulation of biotechnology, along with quantitative tools for measuring uncertainty that forms the basis of quantitative risk assessment, could substantially benefit the governance, oversight, and regulation of future biotechnology products. In addition, early introduction to these concepts might spawn innovation in regulatory science and in product design that takes into account risk-analysis processes.

REFERENCES

EOP (Executive Office of the President). 2015. Memorandum for Heads of Food and Drug Administration, Environmental Protection Agency and Department of Agriculture. July 2. Available at https://obamawhitehouse.archives.gov/sites/default/files/microsites/ostp/modernizing_the_reg_system_for_biotech_products_memo_final.pdf. Accessed January 31, 2017.

EOP. 2016. National Strategy for Modernizing the Regulatory System for Biotechnology Products. Available at https://obama whitehouse.archives.gov/sites/default/files/microsites/ostp/biotech_national_strategy_final.pdf. Accessed January 31, 2017.

EOP. 2017. Modernizing the Regulatory System for Biotechnology Products: An Update to the Coordinated Framework for the Regulation of Biotechnology. Available at https://obamawhitehouse.archives.gov/sites/default/files/microsites/ostp/2017_coordinated_framework_update.pdf. Accessed January 30, 2017.

EPA (U.S. Environmental Protection Agency). 2015. Notice of a Public Meeting and Opportunity for Public Comment on Considerations for Risk Assessment of Genetically Engineered Algae. Federal Register 80:51561–51562.

FWS (U.S. Fish and Wildlife Service). 2011. Charter: Conservation Genetics Community of Practice. October 31. Available at https://www.fws.gov/ConservationGeneticsCOP/pdfs/Charter_COP_10312011.pdf. Accessed January 10, 2017.

Hayes, K.R., B. Leung, R. Thresher, J.M. Dambacher, and G.R. Hosack. 2014. Meeting the challenge of quantitative risk assessment for genetic control techniques: A framework and some methods applied to the common Carp (*Cyprinus carpio*) in Australia. Biological Invasions 16(6):1273–1288.

Ji, X., S.A. Ripp, A.C. Layton, G.S. Sayler, and J.M. DeBruyn. 2013. Assessing long term effects of bioremediation: Soil bacterial communities 14 years after polycylic aromatic hydrocarbon contamination and introduction of a genetically engineered microorganism. Journal of Bioremediation & Biodegradation 4:209.

Murphy, B., C. Jansen, J. Murray, and P. De Barro. 2010. Risk Analysis on the Australian Release of *Aedes aegypti* (L.) (Diptera: Culicidae) Containing *Wolbachia*. Indooroopilly, Australia: CSIRO.

Murray, J.V., C.C. Jansen, and P. De Barro. 2016. Risk associated with the release of *Wolbachia*-infected *Aedes aegypti* mosquitoes into the environment in an effort to control dengue. Frontiers in Public Health 4:43.

NASEM (National Academies of Sciences, Engineering, and Medicine). 2016a. Gene Drives on the Horizon: Advancing Science, Navigating Uncertainty, and Aligning Research with Public Values. Washington, DC: The National Academies Press.

NASEM. 2016b. Genetically Engineered Crops: Experiences and Prospects. Washington, DC: The National Academies Press.

NRC (National Research Council). 2015. Industrialization of Biology: A Roadmap to Accelerate the Advanced Manufacturing of Chemicals. Washington, DC: The National Academies Press.

Trögl, J., A. Chauhan, S. Ripp, A.C. Layton, G. Kuncová, and G.S. Sayler. 2012. *Pseudomonas fluorescens* HK44: Lessons learned from a model whole-cell bioreporter with a broad application history. Sensors 12(2):1544–1571.

Glossary

Allele One of the variant forms of a gene at a particular location (that is, locus) on a chromosome. Different alleles produce variation in inherited characteristics, such as blood type.

Bioeconomy Research and innovation in the biological sciences used to create economic activity and public benefit.

Biotechnology A number of methods other than selective breeding and sexually crossing organisms to endow new characteristics in organisms.

Biotechnology product A product developed through genetic engineering (including products where the engineered DNA molecule is itself the "product," as in an engineered molecule used as a DNA storage medium) or the targeted or in vitro manipulation of genetic information of organisms, including plants, animals, and microbes. The term also covers some products produced by such plants, animals, and microbes or their derived products.

Comparator A known nonbiotechnology organism that is similar to the engineered organism except for the engineered trait.

CRISPR (clustered regularly interspaced short palindromic repeat) A naturally occurring mechanism of immunity to viruses found in bacteria that involves identification and degradation of foreign DNA. This natural mechanism has been manipulated by researchers to develop genome-editing techniques.

***De novo* genome sequencing** Determination of the DNA sequence of the genome (full genetic complement) of an organism.

Dimensionality The spatial and temporal scales of a risk assessment.

Epigenome The physical factors affecting the expression of genes without affecting the actual DNA sequence of the genome.

Expression   The result of a gene being transcribed into RNA, translated into a protein, and ultimately conferring a trait.

Gene drive   A system of biased inheritance in which the ability of a genetic element to pass from a parent to its offspring through sexual reproduction is enhanced. Thus, the result of a gene drive is the preferential increase of a specific *genotype*, the genetic makeup of an organism that determines a specific *phenotype* (trait), from one generation to the next, and potentially throughout the population.

Genetic engineering   Introduction or change of DNA, RNA, or proteins by human manipulation to effect a change in an organism's genome or epigenome.

Genome   The complete sequence of the DNA in an organism.

Genome editing   Specific modification of the DNA of an organism to create mutations or introduce new alleles or new genes; used interchangeably with the term *gene editing.*

Genome engineering   The use of tools that allow rapid and precise changes directly across chromosomes of living cells instead of limiting modifications at single genes.

Genomics   The study of the genome which typically involves sequencing the genome and identifying genes and their functions.

Genotype   The genetic identity of an individual. Genotype often is evident by outward characteristics.

Governance   A set of values, norms, processes, and institutions through which society manages technology development and deployment and resolves conflict formally or informally. Governance includes oversight, which is defined more narrowly as watchful and responsible care or regulatory supervision.

Horizon scanning   A technique for detecting early signs of potentially important developments through a systematic examination of potential threats and opportunities, with emphasis on new technology and its effects on the issue at hand.

Interested and affected parties   People, groups, or organizations that decide to become informed about and involved in a risk characterization or decision-making process. Interested parties may or may not be affected parties, who are people, groups, or organizations that may experience benefit or harm as the results of a hazard or of the processing leading to a decision about risk.

Mesocosm   A bounded and partially enclosed outdoor experimental unit that closely simulates the natural environment.

Metabolomics   Systematic global analysis of nonpeptide small molecules, such as vitamins, sugars, hormones, fatty acids, and other metabolites. It is distinct from traditional analyses that target only individual metabolites or pathways.

Nontarget effects   Unintended, short- or long-term consequences for one or more organisms *other than* the organism intended to be affected by an action or intervention. Concern about nontarget effects typically centers around unforeseen harms to other species or environments, but nontarget effects can also be neutral or beneficial.

Off-target effects   Unintended, short- or long-term consequences of an intervention on the genome of the organism in which the intended effect was incorporated.

Oversight   See *Governance.*

Phenotype/Phenotypic   The visible and/or measurable characteristics of an organism (i.e., how it appears outwardly and physiologically) as opposed to its genotype, or genetic characteristics.

Problem formulation   The scoping phase of a risk assessment, in which the characteristics and use pattern of the product to be assessed are documented, as are the ecosystem or human population potentially at risk and the endpoints that will be the focus of the assessment.

Reagent   Generally a chemical used in a science experiment; in the context of genome editing, a chemical that is used to modify DNA.

Recombinant DNA   A novel DNA sequence created by joining DNA molecules that are not found together in nature.

Regulation   A subcategory of oversight and governance that represents an authoritative rule dealing with details or procedure or a rule or order issued by an executive authority or regulatory agency of a government and having the force of law.

Regulatory science   The development and implementation of risk-analysis methods and the maximization the utility of risk analyses to inform regulatory decisions for biotechnology products, consistent with human health and environmental risk–benefit standards provided in relevant government statutes. Regulatory science includes establishment of information and data quality standards, study guidelines, and generation of data and information to support risk analyses. It can include the development of risk-mitigation measures as well as the development and implementation of safety training and certification programs to help ensure the intended benefits of products are realized and risks to workers, users, and the environment are minimized. Individuals in government, industry, academia, and nongovernmental organizations that contribute to the advancement of regulatory science have degrees across disciplines in the natural, socioeconomic, and computational sciences, engineering, and public policy.

Risk analysis   Risk assessment, risk communication, risk management, and policy relating to risk to human health and the environment, in the context of risks of concern to individuals; to public, private, and nongovernmental organizations; and to society at a local, regional, national, or global level.

Risk-assessment endpoint   Societal, human health, or environmental values that need to be managed or protected.

RNA interference (RNAi)   A natural mechanism found in nearly all organisms in which the levels of transcripts are reduced or suppressed and can be exploited with biotechnology to modify an organism.

Stressor   Any agent or actor with the potential to alter a component of the ecosystem.

Synthetic biology   The application of engineering principles to reduce genetics into DNA "parts" so that those parts can be understood in terms of how they can be combined to build desired functions in living cells. Through this process, it is possible to assemble new organisms from parts of DNA from more than one source organism or to build synthetic DNA from molecules.

Trait   A genetically determined characteristic or condition.

Transcriptomics   The study of transcripts including the number, type, and modification, many of which can impact phenotype.

Transgene   Any gene transferred into an organism by genetic engineering.

Transgenic organism   An organism that has had genes that contain sequences from another species or synthetic sequences introduced into its genome by genetic engineering.

Appendix A

Biographical Sketches of Committee Members

Richard M. Murray (*Chair*) is Thomas E. and Doris Everhart Professor of Control and Dynamical Systems and Bioengineering at California Institute of Technology. He received his BS degree in electrical engineering from California Institute of Technology in 1985 and his MS and PhD degrees in electrical engineering and computer sciences from the University of California, Berkeley, in 1988 and 1991, respectively. Professor Murray's research is in the application of feedback and control to mechanical, information, and biological systems. Current projects include integration of control, communications, and computer science in multiagent systems, information dynamics in networked feedback systems, analysis of insect flight control systems, and biological circuit design. Professor Murray has recently developed a new course at Caltech that is aimed at teaching the principles and tools of control to a broader audience of scientists and engineers, with particular emphasis on applications in biology and computer science. Professor Murray is co-founder and board member of Synvitrobio, a startup biotechnology company focused on commercialization of cell-free synthesis methods.

Richard M. Amasino is a professor with the Department of Biochemistry at the University of Wisconsin–Madison. His work focuses on how plants perceive seasonal cues such as changing day length and temperature and how they use such cues to determine when to initiate flowering. His most recent focus has been on understanding the biochemical pathway through which perception of winter cold leads to flowering in the spring—a process known as vernalization. Dr. Amasino is also a member of the Great Lakes Bioenergy Research Center, which is one of the three bioenergy research centers established by the U.S. Department of Energy. His work with the center involves studying the biochemical basis of plant biomass accumulation as well as directing the education and outreach program of the center. Dr. Amasino is a Howard Hughes Medical Institute professor, a member of the U.S. National Academy of Sciences, and a fellow of the American Association for the Advancement of Science. His teaching and research have resulted in several national and international awards, including the Alexander von Humboldt Foundation Award in 1999. He has served both as president and chair of the board of trustees of the American Society of Plant Biologists. Dr. Amasino received his BS in biology from Pennsylvania State University and his MS and PhD in biology/biochemistry from Indiana University.

Steven P. Bradbury is a professor of environmental toxicology in the Departments of Natural Resource Ecology and Management and Entomology. He is also a faculty member in Iowa State University's Graduate Toxicology Program. Dr. Bradbury is contributing to research, teaching, and extension in university-wide toxicology, environmental, agriculture, and natural resource science and policy programs. Areas of emphasis include pesticide resistance management; pollination services and monarch butterfly conservation; and sustainable agriculture, including the role of integrated pest management within nested layers of governance. Dr. Bradbury retired from the U.S. Environmental Protection Agency (EPA) in 2014. During his last 4 years at EPA he was the Director of the Office of Pesticide Programs. In this role he led evaluation of new and existing pesticides, including biotechnology products; led integration of federal pesticide registration decisions within related international, national, state, and stakeholder-initiated programs; and addressed management options for emerging, high-impact pests, pesticide resistance, and water quality, endangered species, and pollinator protection. Prior to joining the pesticide program in 2002, Dr. Bradbury had more than 15 years of experience in EPA's Office of Research and Development leading efforts to advance human health and ecological risk assessments in support of water quality, pesticide, and industrial chemical programs. Dr. Bradbury has a BS in molecular biology from the University of Wisconsin–Madison and an MS in entomology (insecticide toxicology) and a PhD in toxicology and entomology from Iowa State University. He has published more than 70 peer-reviewed journal articles and book chapters. In 2014, Dr. Bradbury received the Henry A. Wallace Award for Outstanding Leadership to National and International Agriculture from the College of Agriculture and Life Sciences, Iowa State University.

Barbara J. Evans joined the University of Houston Law Center, a Texas state-supported educational institution, in 2007 and currently holds the Alumnae College Professorship in Law and is Director of the Center for Biotechnology & Law at the law school. She teaches and conducts research in the areas of data privacy, health information system governance, and legal issues in genomic testing, gene editing, and precision medicine. She was named a Greenwall Foundation Faculty Scholar in Bioethics for 2010–2013 and has been elected to membership in the American Law Institute. Prior to pursuing an academic career, she was a partner in the international regulatory practice of a large New York law firm and also advised clients on U.S. privacy, research, and medical device regulatory matters. From 2004 to 2007, she was a research professor of medicine and director of the Program in Pharmacogenomics, Ethics, and Public Policy at the Indiana University School of Medicine/Center for Bioethics. She holds an electrical engineering degree from The University of Texas at Austin; MS and PhD degrees from Stanford University; a JD from Yale Law School; and an LLM in health law from the University of Houston, and she completed a postdoctoral fellowship in clinical ethics at the M.D. Anderson Cancer Center.

Steven L. Evans is currently a Fellow at Dow AgroSciences in Seeds Discovery R&D. He received his BA and BS degrees in chemistry and microbiology from the University of Mississippi and a PhD in microbial physiology from the University of Mississippi Medical School. He was a National Institutes of Health postdoctoral fellow at the University of California, Berkeley, and subsequently with the U.S. Department of Agriculture (USDA) in Peoria, Illinois. In 1988 he joined Mycogen Corporation, now Dow AgroSciences, where he has been involved in the development of natural and recombinant biopesticides, including several crop traits from the Mycogen pipeline. At USDA and subsequently in industry roles, Dr. Evans blends high-resolution chemical analysis with enzymology to research agricultural applications of biotechnology. He continues to identify and acquire differentiating biotechnology capabilities. Dr. Evans is chair emeritus of the Industrial Advisory Board of the National Science Foundation–sponsored SynBERC synthetic biology consortium, serves on the Executive Board of the nonprofit Engineering Biology Research Consortium, and is co-chair of the Biotechnology Innovation Organization's Industrial and Environmental Section synthetic biology subteam.

Farren Isaacs is an assistant professor of molecular, cellular, and developmental biology at Yale University. He received a BSE degree in bioengineering from the University of Pennsylvania and obtained his PhD from the Biomedical Engineering Department and Bioinformatics Program at Boston University. In his PhD he integrated theory and experiment to study gene regulatory network dynamics and then pioneered the design and development of synthetic RNA components capable of probing and programming cellular function. He then was a research fellow in the Department of Genetics at Harvard Medical School working on genome-engineering technologies with George Church. At Harvard, he developed enabling technologies for genome engineering, including MAGE (Multiple Automated Genome Engineering) and CAGE (Conjugative Assembly Genome Engineering). His research is focused on developing foundational genomic and biomolecular engineering technologies with the goal of developing new genetic codes and engineered cells that serve as factories for chemical, drug, and biofuel production. He has recently been named a "rising young star of science" by Genome Technology Magazine, a Beckman Young Investigator by the Arnold and Mabel Beckman Foundation, and recipient of a Young Professor award from DuPont. Dr. Isaacs is also co-founder and chief technology advisor of enEvolv, a startup biotechnology firm aimed at commercializing the MAGE technology he co-invented.

Martha A. Krebs is senior scientist in The Pennsylvania State University's College of Engineering and principal investigator and director of the Consortium for Building Energy Innovation at The Navy Yard in Philadelphia. In her most recent previous position, Dr. Krebs worked with University of California, Davis (UC Davis), faculty and staff to leverage and expand research programs through federal, state, and private partnerships. In that role she also has served as science advisor for the California Energy Commission. Before joining UC Davis, she was the commission's deputy executive director for research and development (R&D). From 1993 to 2000, Dr. Krebs served as assistant secretary and director of the Office of Science at the U.S. Department of Energy, responsible for the basic research program that supports the department's energy, environmental, and national-security missions. She also advised the Secretary of Energy on the department's R&D portfolio and the institutional health of its national laboratories. From 1983 to 1993, Dr. Krebs served as an associate director for planning and development at the U.S. Department of Energy's Lawrence Berkeley National Laboratory, where she was responsible for strategic planning for research and facilities, technology transfer, and science education and outreach. From 1977 to 1983, she served on the House Committee on Science first as a professional staff member and then as subcommittee staff director, responsible for authorizing the department's non-nuclear energy technologies and energy science programs. Dr. Krebs received her bachelor's degree and doctorate in physics from the Catholic University of America. She is a fellow of the American Physical Society, the American Association for the Advancement of Science, and the Association of Women in Science.

Jennifer Kuzma is the Goodnight-North Carolina GlaxoSmithKline Foundation Distinguished Professor in Social Sciences and co-director of the Genetic Engineering and Society Center at North Carolina State University. Prior to this position she was a faculty member in science and technology policy at the Humphrey School of Public Affairs, University of Minnesota (2003–2013); study director at the National Research Council in Washington, DC, for genetic engineering and bioterrorism (1999–2003); and an American Association for the Advancement of Science Risk

Policy Fellow at the U.S. Department of Agriculture (1997–1999). She has more than 100 scholarly publications on emerging technologies and governance and has been studying genetic engineering and its societal aspects for over 25 years. She discovered the bacteria product isoprene, a precursor to natural rubber, during from her PhD work in biochemistry, and her postdoctoral work in plant molecular biology resulted in a publication in the journal *Science*. Dr. Kuzma serves on several national and international advisory boards, including the World Economic Forum's Global Futures Council on Technology, Values, and Policy; the Scientific Advisory Board of the Center for Science in the Public Interest; and the U.S. Council on Agricultural Science and Technology's Task Force on Gene-Editing. She has held several leadership positions, including the Society for Risk Analysis Council & Secretary, Chair of the Gordon Conference on S&T Policy, the Food and Drug Administration's Blood Products Advisory Committee, and the United Nations WHO-FAO Expert Group for Nanotechnologies in Food and Agriculture. In 2014, she received the Society for Risk Analysis Sigma Xi Distinguished Lecturer Award for recognition of her outstanding contributions to the field of risk analysis. She has been called upon in national media for her expertise on genetic-engineering policy issues, including recently in *The Washington Post*, *Scientific American*, *The New York Times*, 2015 World's Fair exhibit, *Nature*, and National Public Radio.

Mary E. Maxon is the Biosciences Area Principal Deputy at Lawrence Berkeley National Laboratory, where she is responsible for developing strategies for the use of biosciences to address national-scale challenges in energy and environment. Previously, she was Assistant Director for Biological Research at the White House Office of Science and Technology Policy (OSTP) in the Executive Office of the President, where she developed the National Bioeconomy Blueprint. Before moving to OSTP, Dr. Maxon ran the Marine Microbiology Initiative at the Gordon and Betty Moore Foundation, which supports the application of molecular approaches and comprehensive models to detect and validate environmentally induced changes in marine microbial ecosystems. Prior to that, Dr. Maxon served as Deputy Vice Chair at the California Institute for Regenerative Medicine, where she drafted the intellectual property policies for California stem cell grantees in the nonprofit and for-profit research sectors. Previously, she was Associate Director and Anti-infective Program Leader for Cytokinetics, a biotechnology company in South San Francisco and team leader at Microbia, Inc., based in Cambridge, Massachusetts, where she contributed to the discovery and development of the Precision Engineering technology for production of commercial products using metabolic engineering. Dr. Maxon received her PhD from the University of California, Berkeley, in molecular cell biology and did postdoctoral research in biochemistry and genetics at the University of California, San Francisco.

Raul F. Medina's research interests center around the role that ecological factors play in the population genetics of arthropods. Dr. Medina is particularly interested in the incorporation of evolutionary ecology considerations into pest control practices. His laboratory is currently assessing how species interactions at macroscopic (e.g., host–parasite associations) and microscopic (e.g., arthropod microbiomes) levels may affect genetic variation of agricultural pests and arthropod vectors of human disease. Dr. Medina is currently exploring if the same principles governing insect herbivores' adaptation to their hosts translate in arthropod parasites of animals. Dr. Medina completed his bachelor's degree in biology in Lima, Peru, at the Universidad Nacional Agraria La Molina. He then obtained a Graduate Certificate in conservation biology from the University of Missouri in Saint Louis. He received his master's degree and PhD from the University of Maryland working on predation of forest caterpillars and on hymenopteran parasitoid population genetics, respectively. Soon after his PhD, Dr. Medina started working at Texas A&M University, where he is an associate professor.

David Rejeski is the director of the Science, Technology, and Innovation Program at the Environmental Law Institute (ELI), a nonpartisan research institute in Washington, DC. His research at ELI focuses on better understanding the environmental impacts and opportunities created through emerging technologies and their underlying innovation systems, from synthetic biology to 3-D printing; structural changes in the economy driven by sharing platforms, new business models, and financing systems such as crowdfunding; and new roles for the public in environmental protection, provided through citizen science, do-it-yourself biology, makers, or other emergent, distributed networks of people and things. He co-founded the Serious Games movement in 2003 and Games for Change in 2004 (http://www.gamesforchange.org) and is interested in the use of video game technologies to help engage the public around complex system challenges facing policy makers. Prior to ELI, he directed the Science, Technology and Innovation Program at the Woodrow Wilson Center. He also worked at the White House Office of Science and Technology Policy, the Council on Environmental Quality, and the Environmental Protection Agency (Office of Policy, Planning and Evaluation). He is a Fellow of the National Academy of Public Administration, a guest researcher at the International Institute of Applied Systems Analysis in Austria, a member of EPA's National Advisory Council on Environmental Policy and Technology, and a board member of American University's Center on Environmental Policy. He has been a visiting scholar at Yale University's School of Forestry and Environmental Studies and previously served on EPA's Science Advisory Board and the National Science Foundation's Advisory Committee on Environmental Research and Education. He has graduate degrees in public administration and environmental design from Harvard and Yale Universities.

Jeffrey Wolt is a professor in the programs of Agronomy, Environmental Science, and Toxicology at Iowa State University, where he is affiliated with the Biosafety Institute for Genetically Modified Agricultural Products and co-directs the Crop Bioengineering Center. He started his academic career studying biology at Case Western Reserve University and completed his BS in bioagricultural science at Colorado State University. He received his MS and PhD in agriculture from Auburn University with emphasis in environmental soil chemistry. His expertise includes soil solution chemistry, environmental chemistry, biogeochemistry, ecotoxicology, and risk assessment. Prior to coming to Iowa State, he held academic appointments with the University of Tennessee, the University of Hawaii, and Purdue University. He also worked as an environmental chemist and risk analyst with Dow Chemical. Dr. Wolt's current research interests include biotechnology safety analysis applied to risk management and science policy decision making; environmental and ecotoxicological risk assessment; soil and environmental chemistry applied to exposure assessment, efficacy, environmental monitoring, environmental toxicology, and environmental fate of xenobiotics and genetically modified agricultural products; and applied soil solution chemistry. He also works with regulators and scientists throughout the world to formulate and promote harmonized approaches for assessing the safety of genetically engineered plants. His laboratory group works on the environmental fate of plant products introduced into agroecosystems.

Appendix B

Agendas of Information-Gathering Sessions

Information-gathering sessions include in-person, public meetings and webinars held by the committee from April to August 2016. They are listed in chronological order. The locations of in-person meetings are provided. Presentations that were made via the Internet at the in-person, public meetings are noted.

APRIL 18, 2016—FIRST PUBLIC MEETING

The first in-person, public meeting of the Committee on Future Biotechnology Products and Opportunities to Enhance Capabilities of the Biotechnology Regulatory System was held at the National Academy of Sciences building in Washington, DC.

Open Session Agenda
Monday, April 18, 2016
1:00 PM–5:00 PM

1:00 pm   **Welcome and Introduction to the Committee**
Richard Murray, *Chair*
Douglas Friedman, *Study Director*

1:10 pm   **White House Office of Science and Technology Policy**
Robbie Barbero, *Assistant Director for Biological Innovation*

1:30 pm   **U.S. Department of Agriculture–Animal and Plant Health Inspection Service**
John Turner, *Director, Biotechnology Risk Analysis Program*

Lisa Ferguson, *National Director, Policy Permitting and Regulations Services, National Import Export Services*

2:00 pm   **U.S. Environmental Protection Agency**
                 Rebecca Edelstein, *Team Leader for New Chemicals Program, Chemical Control Division, Office of Pollution Prevention and Toxics*

                 Chris Wozniak, *Biotechnology Special Assistant, Biopesticides and Pollution Prevention Division, Office of Pesticide Programs*

2:30 pm   **U.S. Food and Drug Administration**
                 Ritu Nalubola, *Senior Policy Advisor, Office of the Commissioner*

3:00 pm   **Break**

3:20 pm   **Committee Discussion with Sponsors**

4:30 pm   **Public Comment Period**

5:00 pm   **Adjourn Open Session**

JUNE 1–2, 2016—SECOND PUBLIC MEETING

The second in-person, public meeting of the Committee on Future Biotechnology Products and Opportunities to Enhance Capabilities of the Biotechnology Regulatory System was held at the National Academy of Sciences building in Washington, DC.

Open Session Agenda
Wednesday, June 1, 2016
9:00 AM–5:30 PM

9:00 am   **Trends in Biotechnology Funding and Tools**
                 Juan Enriquez, *Excel Venture Management*

                 Discussion About Trends in Biotechnology Investment and Development
                 (Panel discussion; Moderator: Rick Johnson)

                 Lionel Clarke, *U.K. Synthetic Biology Leadership Council*
                 Juan Enriquez, *Excel Venture Management*
                 Theresa Good, *National Science Foundation*
                 Pablo Rabinowicz and Todd Anderson, *U.S. Department of Energy*

10:45 am   **Break**

11:00 am   **Enabling Tools**
                 Discussion About Trends in Tools Enabling Biotechnology Products of the Future
                 (Panel discussion; Moderator: Farren Isaacs)

                 Nathan Hillson, *Joint BioEnergy Institute*
                 Kevin Munnelly, *Gen9*

Brynne Stanton, *Ginkgo Bioworks*
Bill Peck, *Twist Bioscience* (remote)
Rachel Haurwitz, *Caribou Biosciences, Inc.* (remote)

12:45 pm   **Lunch**

1:30 pm   **Risk Framing Considerations**
               (Discussion; Moderator: Richard Murray)
               Ortwin Renn, *Stuttgart University*, Germany (remote)

2:15 pm   **Break**

2:45 pm   **Open-Release Biotechnology**
               Discussion of Future Emerging Open-Release Products
               (Panel discussion; Moderator: Mary Maxon)

               Thomas Reed, *Intrexon*
               Stephen Herrera, *Evolva*
               John Cumbers, *Synbiobeta* (remote)
               Christopher DaCunha, *Universal Biomining*
               Dan Jenkins, *Monsanto*

5:00 pm   **Public Comment Period**

5:30 pm   **Adjourn Open Session**

Open Session Agenda
Thursday, June 2, 2016
9:00 AM–4:30 PM

9:00 am   **Discussion of "Different" Risks of Open-Release Products**
               (Panel discussion; Moderator: Steven Bradbury)

               Norman Ellstrand, *University of California, Riverside*
               Keith Hayes, *CSIRO*, Australia
               Gregory Jaffe, *Center for Science in the Public Interest*
               Terry Medley, *DuPont*
               Doug Gurian-Sherman, *Center for Food Safety*
               David Hanselman, *Synthetic Genomics, Inc.*

11:30 am   **Lunch**

12:30 pm   **Tools and Opportunities to Enhance Risk Analysis**
               (Panel discussion; Moderator: Jeff Wolt)

               Nathan Hillson, *Joint BioEnergy Institute*
               Shengdar Tsai, *Harvard University*
               David Hanselman, *International Gene Synthesis Consortium*

2:00 pm	**Break**

2:30 pm **Implications of Accessible Biotechnology**
(Panel discussion; Moderator: Richard Amasino)

Todd Kuiken, *Woodrow Wilson International Center for Scholars*
Ellen Jorgensen, *GenSpace*
Tom Burkett, *Baltimore Under Ground Science Space*
Randy Rettberg, *iGEM*

4:00 pm **Public Comment Period**

4:30 pm **Adjourn Open Session**

JUNE 27, 2016—THIRD PUBLIC MEETING

The third in-person, public meeting of the Committee on Future Biotechnology Products and Opportunities to Enhance Capabilities of the Biotechnology Regulatory System was held in San Francisco, CA.

Open Session Agenda
Monday, June 27, 2016
8:30 AM–5:30 PM

8:30 am **Welcome and Introductions**
Richard Murray, *Committee Chair*

8:40 am **New Sector Identification**
(Panel discussion; Moderator: Rick Johnson)

David Berry, *Flagship Ventures*
Craig Taylor, *Alloy Ventures*
Karl Handelsman, *Codon Capital*
Ron Shigeta, *IndieBio*

9:50 am **Break**

10:10 am **Horizon Scanning**
(Panel discussion; Moderator: Mary Maxon)

Adam Arkin, *Lawrence Berkeley National Laboratory*
Ron Davis, *Stanford University*
Andrew Hessel, *Autodesk*
Peter Licari, *Terravia*
Patrick Westfall, *Zymergen*

11:30 am **Lunch**

12:30 pm	**Small Business Perspectives** (Panel discussion; Moderator: David Rejeski)

Antony Evans, *TAXA Biotechnologies*
Kevin Jarrell, *Modular Genetics*
Isha Datar, *New Harvest*
Bruce Dannenberg, *Phytonix*
Alicia Jackson, *Drawbridge Health*

1:50 pm	**Break**

2:05 pm	**Potential Risks Associated with Biotechnology in the Environment and Related Tools: Intentionally Released Biotechnologies** (Panel discussion; Moderator: Richard Murray)

Fern Wickson, *GenOk Centre for Biosafety* (remote)
Michael Hansen, *Consumers Union* (remote)
Steven Strauss, *Oregon State University*
Wayne Landis, *Western Washington University*
Anne Kapuscinski, *Dartmouth College* (remote)

3:25 pm	**Break**

3:40 pm	**Biotechnologies Intended for Contained Use** (Panel discussion; Moderator: Jennifer Kuzma)

Vincent Sewalt, *DuPont*
Craig Criddle, *Stanford University*
Rachel Smolker, *Biofuelwatch/Global Justice Ecology Project* (remote)
David Babson, *U.S. Department of Energy, Bioenergy Technologies Office* (*formerly with Union of Concerned Scientists*)

5:00 pm	**Public Comment Period**

5:30 pm	**Adjourn Open Session**

WEBINARS

July 21, 2016	**Webinar: Safeguarding the Bioeconomy** Gigi Kwik Gronvall, *University of Pittsburgh Medical Center, Center for Health Security* Diane DiEuliis, *National Defense University* Edward You, *Federal Bureau of Investigation, Weapons of Mass Destruction Directorate*

July 22, 2016   **Webinar: Re-Envisioning Risk Assessment**
Kristen C. Nelson, *University of Minnesota*
Steve Mashuda, *Earthjustice*
Tichafa Munyikwa, *Syngenta*
Zahra Meghani, *University of Rhode Island*

July 25, 2016   **Webinar: Defense and Intelligence Agency Funding**
Justin Gallivan, *Defense Advanced Research Projects Agency*
John Julias, *Intelligence Advanced Research Projects Activity*

July 28, 2016   **Webinar: Gene Drives**
Kevin Esvelt, *Massachusetts Institute of Technology*
Elizabeth Heitman, *Vanderbilt University, Center for Biomedical Ethics and Society and Co-Chair of National Academies Report* Gene Drives on the Horizon

July 29, 2016   **Webinar: Screening Tools**
John Yates, *The Scripps Research Institute*
John Ryals, *Metabolon*
Dan Schlenk, *University of California, Riverside*

August 1, 2016   **Webinar: Assessing the Environmental Impact of Synthetic Biology**
Chris Warner, *U.S. Army Engineer Research and Development Center*
Jed Eberly, *U.S. Army Engineer Research and Development Center*

August 2, 2016   **Webinar: An Overview of the TSCA Updates**
Lynn Bergeson, *Bergeson & Campbell PC*

August 2, 2016   **Webinar: Synthetic Nature and the Future of Conservation**
Kent Redford, *Archipelago Consulting*

Appendix C

Requests for Information

In an attempt to better ascertain the nature and extent of federal research designed to support risk analyses of biotechnology products, the committee solicited input from relevant agencies through a request for information (RFI). The questions posed through the RFI were derived, in part, from the report *Creating a Research Agenda for Ecological Implications of Synthetic Biology* published in 2014 following two workshops organized by the Massachusetts Institute of Technology Program on Emerging Technologies and the Wilson Center's Synthetic Biology Project (Drinkwater et al., 2014) and a workshop and Delphi study on synthetic-biology governance (Roberts et al., 2015). The committee was interested in programmatic work related to fundamental and applied research efforts that can inform human, animal, and ecological risk assessments and social and economic costs and benefits. Research related to potential risks of future human drugs or medical devices was not included in the committee's statement of task and therefore was not part of this RFI, except to the extent such research may be broadly applicable to other biotechnology products.

REQUEST SENT TO AGENCIES

"Rapid scientific advances are expanding the types of products that can be generated through biotechnology. In response to a request from the White House Office of Science and Technology Policy, the National Academies of Sciences, Engineering, and Medicine have convened a committee of experts to identify the kinds of products that may be produced with biotechnology in the next 10 years. The U.S. regulatory system for biotechnology products was originally designed in the 1980s, so the committee will also provide advice on the scientific capabilities, tools, and expertise that may be necessary to regulate those forthcoming products and on whether potential future products could pose different types of risks relative to existing products and organisms. The committee's report is expected to be released at the end of 2016.

"To help it address its statement of task, the Academies committee is requesting information on the status of federal research programs that address future biotechnology products. The questions below are derived, in part, from the report *Creating a Research Agenda for Ecological Implications of Synthetic Biology*, published in 2014 following two workshops organized by the Massachusetts

Institute of Technology Program on Emerging Technologies and the Wilson Center's Synthetic Biology Project. The committee is interested in programmatic work related to fundamental and applied research efforts that can inform human, animal, and ecological risk assessments and socio-economic costs and benefits. Research related to potential risks of future human drugs or medical devices is not included in the committee's statement of task and is therefore not part of this request for information (RFI), except to the extent such research may be broadly applicable to other biotechnology products.

"This RFI addresses the level of intramural and extramural research investments for fiscal years 2012, 2013, 2014, 2015, and 2016 (current and anticipated obligations). It also includes a general question concerning research planning and processes whereby research products are adapted and vetted for use in future risk analyses."

Requested Information

"Please provide an estimate of intramural Full-Time Employees and extramural obligations by fiscal year (FY2012, FY2013, FY2014, FY2015, and FY2016) in each of the areas outlined below. As available, please provide links to associated project descriptions and links to any peer-reviewed publications.

a. **Comparators.** Research addressing the nature and extent to which comparisons of future modified organisms, or communities of modified organisms (including those associated with the human microbiome), can be made to wild-type organisms or communities of organisms, to inform problem formulation, risk characterization, and post-market monitoring and surveillance. Include research to address scenarios where there are no "present-day" analogues to the modified organisms.

b. **Nontarget Gene Effects and Phenotypic Characterization.** Research addressing techniques to assess the nature and extent of effects on nontarget genes and unintended phenotypes; understanding phenotypic functions of new traits and how the environment influences expression of the functions; phenotypic characteristics most relevant to near-term perturbations versus long-term consequences in humans, other organisms, communities, or ecosystems.

c. **Impacts on Nontarget Organisms.** Research addressing exposure of future biotechnology products to humans and other nontarget organisms and resultant toxicity (including allergenic responses). Research addressing changes in nontarget species' populations through indirect effects of future biotechnology products due to perturbations in trophic relationships (e.g., reductions in prey and other food sources) and habitat alteration.

d. **Fitness, Genetic Stability, and Lateral Gene Transfer.** Research addressing approaches to assess gene persistence and stability of genetic material across generations; potential for genes to transfer to unrelated species with increased consistency and reliability.

e. **Control of Organismal Traits.** Research addressing intrinsic and external control measures designed to meet specified levels of risk mitigation for intentional or accidental releases.

f. **Life-Cycle Analyses.** Research on the effects future biotechnology products may have on life-cycle processes, such as water utilization and fossil fuel and mineral extraction and consumption.

g. **Monitoring and Surveillance.** Research addressing options for indicators and spatial and temporal sampling designs for human subpopulations, animals, and ecosystems for broad-based detection capabilities or specific applications in proactive or reactive situations.

h. **Modeling.** Research on the use of conceptual models (e.g., in the problem formulation step of risk assessments), physical models (e.g., human organs on a chip, mesocoms), and computational models to help inform risk-based hypotheses in assessments, to direct collection of additional data to reduce uncertainties in assessments, or to provide definitive findings or predictions in risk characterization.

i. **Economic Costs and Benefits.** Research on techniques to quantify the near-term and long-term economic costs and benefits of future products, including comparative analyses with extant products, that address household, community, regional, national, and international scales.

j. **Social Costs and Benefits.** Research on techniques to quantify the near-term and long-term social costs and benefits of future products, including comparative analyses with extant products, that address household, community, regional, national, and international scales.

k. **Other Areas of Research** not addressed above.

l. **Please describe research planning processes within your organization and with sister federal agencies.** To what extent do risk assessors, risk managers, grant project officers, intramural researchers, and, as appropriate, extramural researchers, meet on a regular basis to identify future research needs and identify steps by which anticipated research products will be adapted and vetted for use in risk assessments or socioeconomic cost–benefit analyses? For example, are there ongoing discussions concerning the nature and extent of potential monitoring and/or surveillance needs for the future? Are there ongoing discussions concerning approaches for standardizing methods for data collection and management?"

RESPONSES FROM AGENCIES

Agency Name	Responded with Information	Responded But No Material to Submit	No Response
Air Force Office of Scientific Research U.S. Air Force		X	
Air Force Research Laboratory			X
Army Research Laboratory		X	
Army Research Laboratory Institute for Collaborative Biotechnologies	X		
Defense Advanced Research Projects Agency		X	
Defense Threat Reduction Agency	X		
Intelligence Advanced Research Projects Activity	X		
National Institute of Standards and Technology			X
National Oceanic and Atmosphere Administration Northwest Fisheries Science Center			X

Agency Name	Responded with Information	Responded But No Material to Submit	No Response
National Oceanic and Atmosphere Administration Southwest Fisheries Science Center			X
National Science Foundation Chemical, Bioengineering, Environmental, & Transport Systems	X		
National Science Foundation Division of Environmental Biology			X
National Science Foundation Division of Industrial Innovation & Partnerships	X		
National Science Foundation Division of Molecular & Cellular Biosciences			X
National Science Foundation Division of Social and Economic Sciences	X		
National Science Foundation Office of Emerging Frontiers			X
National Science Foundation Office of Emerging Frontiers in Research & Innovation			X
Office of Naval Research	X		
U.S. Army Corps of Engineers	X		
U.S. Department of Agriculture	X		
U.S. Department of Energy Office of Biological and Environmental Research	X		
U.S. Department of Energy Office of Energy Efficiency & Renewable Energy			X
U.S. Department of the Interior National Invasive Species Council	X		
U.S. Environmental Protection Agency Office of Pesticide Programs	X		
U.S. Environmental Protection Agency Office of Pollution Prevention and Toxics		X	
U.S. Fish and Wildlife Service			X
U.S. Food and Drug Administration		X	
U.S. Geological Survey			X

REFERENCES

Drinkwater, K., T. Kuiken, S. Lightfoot, J. McNamara, and K. Oye. 2014. Creating a Research Agenda for the Ecological Implications of Synthetic Biology. Joint Workshops by the MIT Program on Emerging Technologies and the Wilson Center's Synthetic Biology Project. Available at https://www.wilsoncenter.org/sites/default/files/SYNBIO_create%20an%20agenda_v4.pdf. Accessed August 10, 2016.

Roberts, J.P., S. Stauffer, C. Cummings, and J. Kuzma. 2015. Synthetic Biology Governance: Delphi Study Workshop Report. GES Center Report No. 2015.2. Available at https://research.ncsu.edu/ges/files/2014/04/Sloan-Workshop-Report-final-ss-081315-1.pdf. Accessed October 10, 2016.

Appendix D

Congressionally Defined Product Categories That the U.S. Food and Drug Administration Regulates

Cosmetic	21 U.S.C. § 321(i)—The term "cosmetic" means:
	Articles intended to be rubbed, poured, sprinkled, or sprayed on, introduced into, or otherwise applied to the human body or any part thereof for cleansing, beautifying, promoting attractiveness, or altering the appearance, and articles intended for use as a component of any such articles; except that such term shall not include soap.
Food	21 U.S.C. § 321(f)—The term "food" means:
	Articles used for food or drink for man or other animals, chewing gum, and articles used for components of any such article.
Food Additive	21 U.S.C. §321(s)—The term "food additive" means:
	Any substance the intended use of which results or may reasonably be expected to result, directly or indirectly, in its becoming a component or otherwise affecting the characteristics of any food (including any substance intended for use in producing, manufacturing, packing, processing, preparing, treating, packaging, transporting, or holding food), if such substance is not generally recognized, among experts qualified by scientific training and experience to evaluate its safety, as having been adequately shown through scientific procedures . . . to be safe under the conditions of its intended use; except that such term does not include [pesticide chemicals and residues, color additives, new animal drugs, or dietary supplements].

Major Food Allergen	21 U.S.C. § 321(qq)—The term "major food allergen" means any of the following:

Milk, egg, fish (e.g., bass, flounder, or cod), Crustacean shellfish (e.g., crab, lobster, or shrimp), tree nuts (e.g., almonds, pecans, or walnuts), wheat, peanuts, and soybeans.

A food ingredient that contains protein derived from a food specified in paragraph 1, except for [specified exceptions].

Dietary Supplement	21 U.S.C. 321(ff)—The term "dietary supplement" means:

1. A product (other than tobacco) intended to supplement the diet that bears or contains one or more of the following dietary ingredients:
   a. a vitamin
   b. a mineral
   c. an herb or other botanical
   d. an amino acid
   e. a dietary substance for use by man to supplement the diet by increasing the total dietary intake or
   f. a concentrate, metabolite, constituent, extract, or combination of any ingredient described in clause a, b, c, d, or e
2. A product that:
   a. is intended for ingestion
   b. is not represented for use as a conventional food or as a sole item of a meal or the diet; and
   c. is labeled as a dietary supplement; and
3. Does not include:
   a. an article that is approved as a new drug, [or] certified as an antibiotic [or] licensed as a biologic

Except for purposes of paragraph g [defining drugs] and section 350f [defining reportable foods for which there is a reasonable probability that use or exposure will cause serious adverse health consequences or death to humans or animals], a dietary supplement shall be deemed to be a food.

Medical Foods	21 U.S.C. § 360ee(b)(3)

The U.S. Food and Drug Administration (FDA) originally created this category administratively by interpreting powers elsewhere provided in its statutes (Hutt et al., 2014), but the Orphan Drug Act Amendments of 1988 later defined it as "a food that is formulated to be consumed or administered enterically under the supervision of a physician and which is intended for the specific dietary management of a disease or condition for which distinctive nutritional requirements, based on scientific principles, are established by medical evaluation."

Color Additive   21 U.S.C. § 321(t)(1)—The term "color additive" means a material which:

1. Is a dye, pigment, or other substance made by a process of synthesis or similar artifice, or extracted, isolated, or otherwise derived, with or without intermediate or final change of identity, from a vegetable, animal, mineral, or other source, and
2. When added or applied to a food, drug, or cosmetic, or to the human body or any part thereof, is capable (alone or through reaction with other substance) of imparting color thereto; except that such term does not include any material which the Secretary, by regulation, determines is used (or intended to be used) solely for a purpose or purposes other than coloring.

Infant Formula   21 U.S.C. § 321(z)—The term "infant formula" means:

A food which purports to be or is represented for special dietary use solely as a food for infants by reason of its simulation of human milk or its suitability as a complete or partial substitute for human milk.

New Animal Drug   21 U.S.C. § 321(v)—The term "new animal drug" means:

Any drug intended for use for animals other than man, including any drug intended for use in animal feed but not including such animal feed

1. The composition of which is such that such drug is not generally recognized, among experts qualified by scientific training and experience to evaluate the safety and effectiveness of animal drugs, as safe and effective for use under the conditions prescribed, recommended, or suggested in the labeling thereof
2. The composition of which is such that such drug, as a result of investigations to determine its safety and effectiveness for use under such conditions, has become so recognized but which has not, otherwise than in such investigations, been used to a material extent or for a material time under such conditions

Provided that any drug intended for minor use or use in a minor species that is not the subject of a final regulation published by the Secretary through notice and comment rulemaking finding that the criteria of paragraphs 1 and 2 have not been met (or that the exception to the criterion in paragraph 1 has been met) is a new animal drug.

Animal Food   21 U.S.C. § 321(f)—See term "food" on p. 209.

Tobacco Product   21 U.S.C. § 321(rr)—The term "tobacco product" means:

Any product made or derived from tobacco that is intended for human consumption, including any component, part, or accessory of a tobacco product (except for raw materials other than tobacco used in manufacturing a component, part, or accessory of a tobacco product).

Drug   21 U.S.C. § 321(g)(1)—The term "drug" means:
(relevant insofar as
it sets boundaries   1.   Articles recognized in the official United States Pharmacopoeia,
on other product   official Homoeopathic Pharmacopoeia of the United States, or
definitions)   official National Formulary, or any supplement to any of them; and
                      2.   Articles intended for use in the diagnosis, cure, mitigation,
                           treatment, or prevention of disease in man or other animals; and
                      3.   Articles (other than food) intended to affect the structure or any
                           function of the body of man or other animals; and
                      4.   Articles intended for use as a component of any article specified in
                           clause 1, 2, or 3

                      A food or dietary supplement [that makes nutritional or health related
                      claims that comply with FDA's regulations at 21 U.S.C. § 353(r)] is not
                      a drug solely because the label or the labeling contains such a claim.
                      A food, dietary ingredient, or dietary supplement for which a truthful
                      and not misleading statement is made in accordance with [21 U.S.C.
                      § 353(r)(6), which requires manufacturers of dietary supplements to
                      substantiate claims and provide certain disclosures to customers] is not
                      a drug.

Device   21 U.S.C. § 321(h)—The term "device" means:
(relevant insofar as
it sets boundaries   An instrument, apparatus, implement, machine, contrivance, implant,
on other product   in vitro reagent, or other similar or related article, including any
categories)   component, part, or accessory, which is

                      1.   Recognized in the official National Formulary, or the United States
                           Pharmacopeia, or any supplement to them,
                      2.   Intended for use in the diagnosis of disease or other conditions, or
                           in the cure, mitigation, treatment, or prevention of disease, in man
                           or other animals, or
                      3.   Intended to affect the structure or any function of the body of man
                           or other animals, and which does not achieve its primary intended
                           purposes through chemical action within or on the body of man or
                           other animals and which is not dependent upon being metabolized
                           for the achievement of its primary intended purposes.

REFERENCE

Hutt, P.B., R.A. Merrill, and L.A. Grossman. 2014. Food and Drug Law: Cases and Materials, 4th Ed. St. Paul, MN:
    Foundation Press.